Lecture Notes in Energy (LNE) is a series that reports on new developments in the study of energy: from science and engineering to the analysis of energy policy. The series' scope includes but is not limited to, renewable and green energy, nuclear, fossil fuels and carbon capture, energy systems, energy storage and harvesting, batteries and fuel cells, power systems, energy efficiency, energy in buildings, energy policy, as well as energy-related topics in economics, management and transportation. Books published in LNE are original and timely and bridge between advanced textbooks and the forefront of research. Readers of LNE include postgraduate students and nonspecialist researchers wishing to gain an accessible introduction to a field of research as well as professionals and researchers with a need for an up-to-date reference book on a well-defined topic. The series publishes single- and multi-authored volumes as well as advanced textbooks. **Indexed in Scopus and EI Compendex** The Springer Energy board welcomes your book proposal. Please get in touch with the series via Anthony Doyle, Senior Editor, Springer (anthony.doyle@springer.com)

More information about this series at http://www.springer.com/series/8874

R.W. Bentley

Introduction to Peak Oil

Springer

R.W. Bentley
Department of Cybernetics
University of Reading
Reading
UK

ISSN 2195-1284 ISSN 2195-1292 (electronic)
Lecture Notes in Energy
ISBN 978-3-319-26370-0 ISBN 978-3-319-26372-4 (eBook)
DOI 10.1007/978-3-319-26372-4

Library of Congress Control Number: 2015956376

Printed on acid-free paper

This Springer imprint is published by SpringerNature
The registered company is Springer International Publishing AG Switzerland

Acknowledgments

This book could not have been written without the help and wise counsel from many people over the years, for which I am extraordinarily grateful. In particular I would like to thank:

In Europe: John Burton, Colin Campbell, Jean Laherrère, George Leckie, Ken Chew, Francis Harper, Richard Miller, Max Coleman, Bruce Sellwood, George Whitfield, Peter Dunn, Roger Booth, Chris Skrebowski, David Fleming, Rayner Mayer, Saad Mannan, Simon Wheeler, Lado Kurdgelashvili, Michael Smith, David Strahan, Richard Hardman, Jeremy Leggett, Peter Jackson, Steve Sorrell, Jamie Speirs, Richard O'Rourke, Hans DuMoulin, Michael Jefferson, John Busby, Chris Vernon, Godfrey Boyle, Pierre-René Bauquis, Jean Marie Bourdaire, Fatih Birol, Chris Besson, Peter Gerling, Klaus Illum, Jens Junghans, Kjell Aleklett, Mikael Höök, Kristofer Jakobsson, Bengt Söderbergh, Rui Rosa, Ugo Bardi, Susanne Peters, Sam Carmalt, Jörg Schindler, Werner Zittel, Rembrandt Koppelaar, Pedro Prieto and Luis De Sousa; in Australia and the Far East: Bruce Robinson, Matt Mushalik, Lianyong Feng, Q. Meng, Xu Tang and Jianliang Wang; and in the USA: Ron Swenson, L. Ivanhoe, Tom Alhbrandt, Gregory Ulmeshek, Ron Charpentier, Robert Hirsch, Mike Rodgers, David Greene, Ken Deffeyes, Charles Hall, Steve Andrews, David Lawrence and Hilary Smith.

In addition, I have much appreciated useful conversations with the following, though often disagreeing with the views expressed: Georges DuPont-Roc, Peter Davies, David Frowd, Hans Maters, Peter Odell, Alex Kemp, Paul Stevens, John Chesshire, John Mitchell, Robert Mabro, M. Adelman and Michael Lynch.

Particular thanks go to David Astor and Sarah Astor who, at David Fleming's suggestion, employed me at ODAC for a year to study peak oil and who kindly arranged for the purchase, at Colin Campbell's suggestion, of the invaluable IHS Energy PEPS '2P' dataset without which much of the research described here could not have been carried out.

The above is only a very partial list of those to whom debts are owed, and many more people have kindly provided invaluable data and advice over the years; my apologies to those I have omitted here.

Finally, heartfelt thanks are due to my wife, Yongmei Bentley, who has tolerantly borne far too many years of 'oil study'.

Drafts of this book have been reviewed at various stages by Colin Campbell, Jean Laherrère, Richard Miller, Michael Smith, Steve Andrews, Sam Carmalt and Charles Hall, for which many thanks are due. The errors that remain are mine, and these are almost certainly many. Comments and corrections are therefore very welcome.

Please note that this is not intended to be a general book on peak oil. Though none of the topics involved is difficult, the book is for those who wish to understand the subject from a technical point of view. The text can be supplemented by any of a number of excellent books on the subject, including Kjell Aleklett's (2012) *Peeking at Peak Oil* (which has a new edition currently in preparation), and Charles Hall & Carlos Ramírez-Pascualli's (2013) *The First Half of the Age of Oil*.

September 2015 R.W. Bentley

Contents

Chapter 1
Introduction

1.1 The Purpose of this Book

For some years now the process of supplying the world with oil has been getting more difficult. Global production of *conventional oil* (the normal oil found in oil fields) is close to its *resource-limited maximum*, and to meet additional demand the world has had to turn increasingly to the production of non-conventional oils, such as tar sands oil and shale ('light tight') oil. This transition has led to a significant rise in the price of oil, and will probably at some point, possibly fairly soon, lead to an overall decline in global oil supply. The aim of this book is to explain the underlying causes for this fundamental change in oil supply, and also why the change is still not widely understood.

1.2 The Importance of Oil, and Hence the Difficulty of Transition

We start however by considering the importance of oil, and hence why a transition in its supply will not be easy.

Oil is a lifeblood of the modern economy, and it makes up the largest part of the world's traded energy, some one-third of the total. In particular oil fuels nearly all the extensive, cheap transport that is a major contributor to today's efficient global economic activity, whether in the production of food or goods, or provision of the services that we rely on. As most people appreciate, if oil is not available in the short term food does not get to the shops, nor workers to their jobs, and society is at risk. This was amply demonstrated during the European 'fuel protests' of 2000.

However the problems we face as global oil supply becomes increasingly difficult are somewhat different. This is because, first, oil can be substituted by other forms of energy in almost any activity. Aeroplanes, for example, can fly on biofuels

R.W. Bentley, *Introduction to Peak Oil*, Lecture Notes in Energy 34,
DOI 10.1007/978-3-319-26372-4_1

or liquid hydrogen; trucks be run on compressed natural gas; cars on gas or electricity; houses and offices be heated and cooled by electricity or by other fuels; and industrial chemicals produced from gas, coal or biomass, or directly from solar energy using ubiquitous feedstocks. Second, given sufficient warning, society can choose to change its activities and priorities so as to use less oil.

But several factors make the current transition to a world of expensive oil, and in time to a world of less oil, both difficult and painful.

The first is simply the very large quantity that we use. As mentioned, oil is currently the world's largest traded energy, and finding adequate substitutes to allow the world to replace this amount of energy over any medium timeframe looks technically very difficult, and also expensive in terms of the investment required. For example, the quantity of currently available biofuels falls far short of what is needed to replace oil; the global supply of *conventional* gas will see its own resource restrictions fairly soon; both coal and nuclear face significant constraints on increased production; and while the *potential resource* of solar energy is more than adequate, transitioning to this on a large scale has severe restrictions of cost and energy return when compared to the energy supplied by oil.

A second factor to consider in the coming energy transition is the impact of the price of oil on economic activity. For the half-century from 1920 to 1970 oil was very cheap, averaging about $15/barrel ('$/bbl') in today's money (its 'real-terms' price), and as a result the global use of oil expanded rapidly, and economic wealth in parallel. Later, for the nearly two decades between 1986 and 2003, the average real-terms price was about double, at $30/bbl, and the growth of oil use was slower, but still substantial. Only for the two periods from 1974 to 1985, and since 2004, has the oil price been significantly higher, reaching at times a real-terms price of $100/bbl or more. In both these periods the high price has resulted in considerable damage to the world's economy. Going forward the on-average high oil price that results from increasingly difficult supply is expected to continue to impact global economic activity; and hence also the world's ability to fund the increasing energy efficiency, and production of new energies, that will be required.

A third factor that makes transition away from oil likely to be difficult is the lower energy return of many of the alternatives. Energy return compares the energy generated by an energy source to that required to produce this energy. For quite a number of the alternatives to oil their energy returns are considerably less than that for oil.

An important final factor to consider is CO_2 emissions. While the oil we use today is no angel in this regard, some of the alternative fuels emit greater levels of CO_2 per unit of energy generated. As seems now well established, a significant percentage of all the world's oil, gas and coal must be left in the ground if the global limit of 2 °C warming is to be avoided.

This book does not look, however, at these broader issues of global all-energy supply; nor at what energy transition rates are possible; and nor at the likely corresponding levels of economic activity. These topics are complex, and require detailed modelling, and also generally better data than are currently available (particularly on energy return by energy source, and hence on corresponding net-energy limits to transition rate) to have a clear view of how the world's energy future can play out.

Instead this book looks at the simpler question of the future availability of oil, and in particular that of 'conventional oil'.

1.3 Conventional Oil Versus Non-conventional Oil

In understanding the rise in oil price since 2004, and also the limits to future oil supply, an important distinction to make is between the production of *conventional* oil and that of *non-conventional* oil.

Oil exists in many forms. It can be found at the land surface or on the seabed as oil seeps; in degraded form in tar pits and in extensive areas of tar sands; as oil's precursor, kerogen, still in the original rock in which it was laid down (and from which it needs retorting to yield 'oil shale' oil); and as light, flowable oil, either still captured in the original rock (as 'shale oil', that needs hydraulic fracturing, 'fracking', to release it); or after having migrated to an open-pored reservoir of rock (an oil field), from which it can be extracted by drilling.

It is this last class of oil, *the relatively light, flowable oil in fields* that is generally classified as conventional oil, and where the bulk of oil production currently, and by far the largest part historically, has been of this class of oil.

By contrast, non-conventional oil tends to be found in extensive regions (within which there may be 'sweet spots'), and where flow to a production well is not possible without significantly changing the nature of the oil itself (for example, by heating to reduce viscosity, addition of a solvent, or retorting), or that of the surrounding material (such as mining the sand in which the oil is contained, or by fracturing the rock in which it is trapped). Non-conventional oil thus includes very heavy oil, oil from tar sands and Venezuela's Orinoco fields, shale ('light tight') oil and oil produced from kerogen by retorting.

Oil, in addition, can be produced from yet other sources. It can come from the physical transformation of some of the gas from gas fields, as either condensate or natural gas liquids ('NGLs'); by chemical transformation of gas from a variety of fossil sources (yielding gas-to-liquids, GTLs), or similarly from coal (coal-to-liquids, CTLs); or alternatively from biomass, either directly as biofuels, or by chemical or biological change from a variety of types of biomass.

Note that NGLs are often included in conventional oil (though in this book we try to break them out separately where possible), while the oil produced from GTLs, CTLs and biomass is often classed as 'other liquids'. Annex 1 gives more detailed definitions.

To see why this distinction between classes of oil is important, we need to ask the following question: Why over a century and a half has the world, in the main, used conventional oil (i.e. oil in fields), rather than oil from the many other sources that exist, and where some of the latter (such as oil from biomass, and from coal and kerogen) were used extensively before conventional oil came to dominate?

The answer is simple: Up to now the oil in fields has usually been far cheaper to produce than these other oils. The reason for this generally lower cost of conventional oil relates principally to flow rate, and energy return.

Flow rate

As noted above, oil in fields is concentrated geographically and flows easily, and hence often yields large flow rates when produced by relatively simple drive mechanisms, such as own pressure, gas drive or water flood.

For example, while the 1859 Drake well, the first commercial oil well in the US, yielded up to about 20 barrels of oil per day ('b/d'), only 2 years later the first major US gusher yielded 4000 b/d, and in 1901 the Spindletop field in Texas flowed at 100,000 b/d. Admittedly in these early years such flows were often short-lived, but subsequent large fields typically have yielded over 500,000 b/d for considerable periods; while the Middle East giants produce 1 million b/d and above, and the world's largest field, Ghawar, averages over 5 million b/d. Thus once located, conventional oil from large oil fields has generally been cheap to produce due to relatively easy production methods and high flow rates.

As a result, while 'the petrol tank in your car does not care' what type of oil (conventional or non-conventional) is used, the user certainly does. The user would far prefer conventional oil at its pre-1973 long-term average real-terms price of \$15/bbl, or even at its post-1985 real-terms average price (up to the 2004 increase) of \$30/bbl, than to have to pay the ~\$60/bbl production cost for US light tight oil,[1] or the more than \$160/bbl for 'Canada oil sand mine upgraded' oil, currently estimated by IHS-CERA (see Fig. 16 of Miller and Sorrell 2014); or the production cost—whatever it will be—of retorted kerogen oil plus carbon capture, or of synthetic fuel made from electrolysis of water plus CO_2.[2]

Energy return

Another way to look at the relative ease of production of conventional oil is in terms of its energy return; nearly all of the non-conventional oils have lower energy returns. Though the data are hard to establish unequivocally, Guilford et al. (2011) and Hall (personal communication) suggest for example that in the US the ratio of energy return to energy invested (EROI) for conventional oil was about 30:1 in the 1930s, rising to 40:1 in the 1970s as scale increased and technology improved, and subsequently falling with production of the more difficult conventional oils, such as deep offshore or Arctic oil, to an average ratio of perhaps 14:1 today. By contrast, nearly all non-conventional oils have lower energy ratios; tar sands, for example, being quoted as having ratios of from 1.5 to 8:1, and corn ethanol as only perhaps 2 or 3:1 (probably higher in Brazil, and in some cases perhaps negative). Since Hall et al. (2009) and Lambert et al. (2014) calculate that modern society will have difficulty in functioning if its fuels have energy ratios of less than perhaps 5–10:1, the current transition from mainly conventional oil to increasing quantities of non-conventional oil is significant, and needs to be understood.

1.4 Oil Reserves Data: Proved Versus Proved-Plus-Probable

Recognising the difference between conventional and non-conventional oil is one part of understanding the current transition in oil supply, but by far the largest reason why this transition has been poorly understood has been confusion over the data, and in particular the reserves data.

The reserves of an oil field or a region estimate the amount of recoverable oil remaining at a given point of time, and the problem lies with the differences between the *proved* oil reserves (the so-called '1P' data), and the *proved-plus-probable* reserves (the '2P' data); see Bentley et al. (2007). The following two sections discuss this topic briefly, and it is discussed more fully in Annex 2.

1.4.1 Proved ('1P') Oil Reserves

Where annual summary tables of oil reserves data are given in the public domain they are usually of proved (1P) reserves, as for example in the BP *Statistical Review of World Energy*, the annual tables in *World Oil* or the *Oil & Gas Journal*, or on the US Energy Information Administration (EIA) website. Although such reserves data are used in company reports they have been extraordinarily misleading on the actual quantity of oil discovered, especially in the past, and they generally cannot be used to forecast oil production despite many analysts having done just this. (For an exception to this rule see Bentley 2015a). The issues with the proved reserves data are as follows:

Understatement

The term 'proved' reserves would seem to indicate solid, reliable data. The BP *Statistical Review* for example quotes such reserves as: '*Generally taken to be those quantities that geological and engineering information indicates with reasonable certainty can be recovered in the future from known reservoirs under existing economic and operating conditions*'.

In fact, this has been far from the case. Historically, proved reserves have been *very much less* than the quantities that 'with reasonable certainty can be recovered', where the latter were far better indicated by the oil industry's proved-plus-probable (2P) reserves data.

In the US and Canada in particular, the apparent size of individual fields, as given by their estimated initial proved reserves, has often grown massively over their lifetimes, typically from five to over ten times their original estimated volumes.

There are many explanations for this, but 'drilling-up' of early fields is likely a major factor, as under SEC rules proved reserves can largely refer only to oil in communication with existing wells. Though the US and Canada have typically shown the largest such proved reserves 'reserves growth', the phenomenon applies to many regions around the world. In the UK, for example, the total oil proved reserves have long stood at only about half the value given by the industry data for 2P reserves.

For *regions*, proved reserves will also be underestimates due to summation. This is because simple addition of any high-probability data, such as proved reserves, significantly underestimates the total at the same probability level.

Naturally, over time, as fields get produced, the conservative proved reserves data generally grow to become close to 2P estimates. This is now the case for many —but by no means all—oil-producing countries.

Overstatement

But the opposite problem also exists, where proved reserves are almost certainly significantly overstated.

Though most countries' public domain proved oil reserves are, as one would expect, smaller than the industry 2P data, in some anomalous but important cases the reverse is true, and the country's published 1P reserves significantly exceed the industry 2P reserves. These cases are mainly the result of the OPEC 'quota wars' step-changes in proved reserves that occurred in the 1980s, when OPEC's allowed production of individual members was based in part on the size of their reserves. This issue has been written about extensively, and may constitute an over-reporting, above the correct 2P reserves, by as much as ~300 Gb, close to a third of remaining conventional oil reserves.

Non-statement

Finally, in the main public-domain 1P databases (e.g. BP *Stats. Review*, or from the *Oil & Gas Journal*), the data on proved reserves are frequently not updated annually, and can remain static for sometimes very long periods of time. (For example, in the end of 2014 *O&GJ* data for 106 oil-producing countries, 66 reported no change in proved reserves during 2014.) For this reason also, such 1P data are significantly in question; and more importantly *trends* in these data are misleading.

1.4.2 Oil Industry Proved-Plus-Probable ('2P') Reserves Data

Now we turn to the proved-plus-probable ('2P') reserves data. Such data can be gathered for individual oil fields from a wide variety of published industry and government sources, but with considerable effort, as has been done for example by Richard Miller, Michael Smith at Globalshift Ltd., and Robelius and others at the University of Uppsala.

Large commercial *by-field* 2P datasets can instead be purchased from firms such as IHS Energy, Wood Mackenzie, PFC Energy and Rystad Energy. Here the data have been assembled and checked, and there is also much additional proprietary information. Such datasets tend to be very expensive. Fortunately, simpler 2P datasets are available at moderate cost. Certain by-field data can be purchased from Globalshift Ltd., and possibly from Peak Oil Consulting, Richard Miller and Uppsala University. In addition, extremely useful *by-country* 2P data are available from IHS Energy's 'PEPS' database, where researchers should use the version with data back to 1834 (but note that US and Canada non-frontier data are only 1P).

Some collected 2P data now are in the public domain. For conventional oil, adjusted global and by-country data are given by Laherrère on websites and elsewhere (including in this book, and as a chart in *The Oil Age,* January 2015); while charts for adjusted 'Regular conventional' oil for a wide range of countries, and including essays on the associated petroleum geology, are given in Campbell (2013). (In the latter, the key graphs are the 'Status of Oil Depletion' plots. These give, by-country, the cumulative backdated 2P discovery data to show how much oil has been found; the country's cumulative production; and also Campbell's judgement of the country's ultimately recoverable resource ('URR'). Evolution over time of a country's 2P reserves, and yet-to-find, are given by the differences between these data.)

Excellent plots of past and forecast production (but not of discovery) for all oil-producing countries, based on detailed by-field 2P data, are free on the Globalshift website (www.globalshift.co.uk). In addition, Rystad Energy's UCubeFree facility gives past and future production by-country, based on their estimates of 2P data.

Comparison of the evolution over time of global 1P oil data with 2P is given in Annex 2. Today, for *conventional* oil, the size of global 1P reserves is roughly the same as that of the 2P reserves. This is because the overstatement of OPEC 1P reserves is roughly matched by the understatement of the 1P reserves in most other countries. Nevertheless, in general 1P reserves data should *not be used* for understanding the future of oil production, and especially not their very misleading evolution over time. The data presented in the graphs and analyses given below are in general *2P data,* drawn from a variety of oil industry datasets.

1.5 Structure of This Book

Now we turn to the structure of this book:

Chapter 2 explains the concept of 'peak oil'. As used in this book, and also fairly generally, the term refers to the maximum in the production of oil that results from the *physical and economic characteristics* of an oil resource. It is shown that such peaks have very different causes, profiles and impacts depending whether they refer to peak of production in a field, in a region, or to the world as a whole. Similar differences also hold true if the peak refers to production of conventional oil only, or to that of 'all-oils' or 'all-liquids'.

Chapter 3 gives a brief history of some of the main forecasts of peak conventional oil production that have been made over the years. Such forecasts can be generated by a variety of approaches, including the 'mid-point' rule, PFC Energy's '60 %' rule, mirroring of production to discovery, Hubbert linearisation, summing logistic curves, or by detailed 'bottom-up' calculations by field.

Chapter 4 then examines oil forecasts from a variety of fairly recent sources to ascertain the likely dates for the global peak of the various types of oil, both conventional and non-conventional. It is shown that in contrast to the widely held view that 'all oil forecasts have been wrong', for over 40 years nearly all oil forecasts that have been based on an understanding of resource limitations have been substantially correct in predicting that the peak in the global production of conventional oil would occur roughly around the year 2000. Much greater attention should have been given to these technically based warnings of significant transition in the global oil supply.

Chapter 5 then turns to the question of why the concept of 'peak oil' has been so poorly understood, both in the past and still today. The primary reasons have been:

- As already mentioned, reliance on the very misleading proved reserves data; and in particular on the apparent *replacement* of these proved reserves over time.
- Not understanding the significance of the 'mid-point peak' in production.
- Reliance on global URR estimates significantly higher than that indicated by the trend of 'proved-plus-probable' (2P) oil discovery.

Chapter 6 sets out some caveats, and then summarises the book's main conclusions. Finally, a number of annexes are given; these give greater detail on topics in the main chapters, and also discuss briefly a number of topics not covered in the main text. The latter include the relationship between peak oil and climate change; peak oil demand; and the global peak of conventional gas production.

Chapter 2
Explaining Peak Oil: What It Is, and Why It Happens

This chapter explains the concept of 'peak oil'. The term, as used in this book and also generally, refers to the point at which the production of oil from an oil field, a region, or the world as whole reaches a maximum and then subsequently declines due primarily to *limitations of resource availability*. Note that there can be several such 'resource-limited' maxima in the production history of a field or a region; in a field for example from the application of new technology or a significant increase in oil price; and likewise in a region, for example from successive phases of discovery.

The term therefore generally does not refer to a peak in production that occurs due to 'above-ground' factors, such as demand reaching a maximum, a country limiting access to the development of its oil, or the imposition of quotas or similar constraints on production.

The physical and economic reasons for a peak in oil production, the shape of the production profile before and after the peak, and usually also the economic significance of the peak, are very different in the case of an individual oil field, a region that contains many fields, and the world as a whole. Moreover, the economic significance of peak will be different if it applies to conventional oil only, or to conventional plus non-conventional oil, or to 'all-liquids'.

Definitions used in this book for different categories of oil, and for extraction techniques, are set out in Annex 1. Importantly, recall from above that *conventional* oil is defined here as essentially *oil in fields*, i.e., primarily mobile oil that has migrated from a source rock to a discrete field (and usually one having an oil-water contact). The reason this definition is adopted is twofold: the generally intrinsically lower cost of this type of oil as already discussed; and because the physical factors that drive production of this oil result in the peak of production in a region as occurring typically when roughly only half of the region's total recoverable oil has been produced; the so-called 'mid-point' peak.

In this chapter and the next we look primarily at the production of *conventional* oil, and look in turn at peak in individual oil fields, in regions, and then in the wold as a whole. We start by examining the production peak in fields.

R.W. Bentley, *Introduction to Peak Oil*, Lecture Notes in Energy 34,
DOI 10.1007/978-3-319-26372-4_2

2.1 The Production Peak in an Individual Oil Field

2.1.1 Typical Field Production Profiles

The production profiles of *individual oil fields* can vary considerably. Figure 2.1 gives examples of fairly typical production profiles for a range of field sizes and locations.

Figure 2.1 includes the 'fairly slowly up and down' production of larger older fields such as Romashkino, but with a very long production tail; the quickly up (then long decline) of relatively modern offshore fields such as Forties, where speed of financial return is important given the high up-front investment; and the relatively long plateaus of both a 'heavily-worked' field like Daqing (and where output was probably judged more important than rate of return), and of a large Middle East

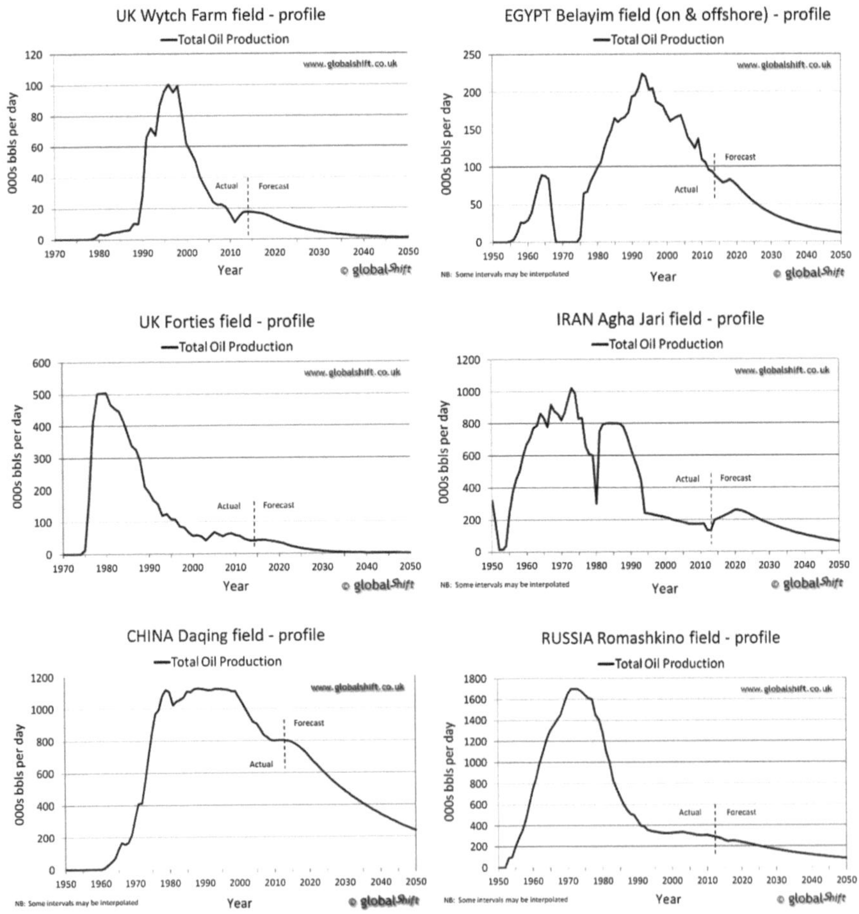

◀ **Fig. 2.1** Oil field production profiles, for a range of field sizes and locations.

Top left Wytch Farm, UKs largest on-shore oil field (though in fact much of the field is offshore, reached by horizontal drilling). Relatively low production until the infrastructure fully in place. Relatively short plateau (at ~100,000 b/d) followed by long steady decline, small recent late recovery. Note the distinction in this plot (and others in this Figure) between actual and forecast data.

Top right Belayim field, one of Egypt's largest fields. This field has both onshore and offshore production zones. Note loss of production due to 1967 war. Relatively symmetric climb to peak (at ~225,000 b/d) and decline (though part of this is forecast).

Centre left Forties, UK's largest field. Rapid rise in production to a short plateau at 500,000 b/d; long decline with some late recovery.

Centre right Aghajari, large Iranian field discovered 1938, in production 1940. Note production fall post-1978, presumably due to the revolution; and expected moderate future recovery from planned gas injection. General profile: relatively long approximate plateau (at up to just over 1 Mb/d) that is typical of large Middle East fields. Field output in decline once it had produced roughly half its likely recoverable oil.

Lower left Daqing, China's largest field. Discovered 1959, in production 1960 with Russian technical assistance. The Chinese are justifiably proud that despite forecasts to the contrary they held production on plateau (at ~1.1 Mb/d) for some 20 years, through the use of water-flood, infill drilling (it is a braided channel field with many separate sandstone reservoirs), and latterly chemical additives. But the field is now well into decline.

Lower right Romashkino, one of Russia's largest fields, discovered in 1948. Roughly symmetric production rise (to ~1.7 Mb/d) and decline, but with a long production 'tail'.

Source Globalshift Ltd.; 2015

field, Aghajari, where supply was constrained over many years for a variety of reasons (for a general explanation for long-term global oil supply constraints see Bentley and Bentley 2015a, b). For other examples of field production profiles, the 'regional by-field' plots of Figs. 2.7, 2.10, 2.13, 2.18 and 2.19 below include the profiles for over a hundred fields.

As these examples indicate, the production profiles of the majority of all fields show production rising fairly early to peak (or a short plateau), and then a long period of decline. Note that the production rate at peak or plateau in part reflects the size of the pipeline and other infrastructure taking oil from the field, and this, in turn, is optimised to maximize field profit over time, which includes consideration of future expected oil price. Note also that in recent years both primary and secondary recovery techniques are often employed from the outset.

For all classes of fields production eventually declines, driven usually by a combination of physical constraints. These include loss of field pressure, reduction in oil volume for wells to access (set by the reducing length of the oil column), increasing water cut if water-drive is used, and by increasing drive-fluid bypass of the oil within the reservoir (where the drive fluid may be naturally occurring, or injected). In turn these factors reflect the specific characteristics of the production techniques use, and of the reservoir itself where the latter include the rock-oil or oil-water interface characteristics, and, importantly, inhomogeneities within the reservoir. As can be seen in Fig. 2.1 and the other Figures quoted, *production peaks in fields when typically somewhere between about a quarter to a half* of the field's recoverable oil has been produced.

2.1.2 Other Field Production Profiles

However not all fields show this 'typical' profile, and two significant exceptions need attention. These are:

(i) Old and very old fields, especially if difficult to produce in some way

Old and very old fields, and especially those difficult to produce, often have distinct production phases, and multiple peaks, as different types of production technology are used across the field's life. One such field is the Kern River heavy oil field in California, discovered in 1899 but which did not peak until the 1970s; see Fig. 2.2, left. (Note that by the definition used in this book, Kern River counts as a non-conventional oil field as its heavy oil—at least in the later stages of production—requires heating to reduce viscosity.) Another such field with an atypical production profile is Weyburn in Saskatchewan, discovered in 1955, which illustrates the large gains sometimes possible from the application of enhanced oil recovery (EOR) techniques; Fig. 2.2, right.

These two fields are often quoted by 'peak oil' sceptics to show the large gains in yield that technology can achieve, and hence to illustrate—in their view—the intrinsic error of forecasts that assume relatively fixed volumes of recoverable oil. Kern River and Weyburn are quoted, for example, in Mills (2008), pp. 84 and 87 respectively; Kern River in Maugeri's comment in *Science* (2004); while Weyburn

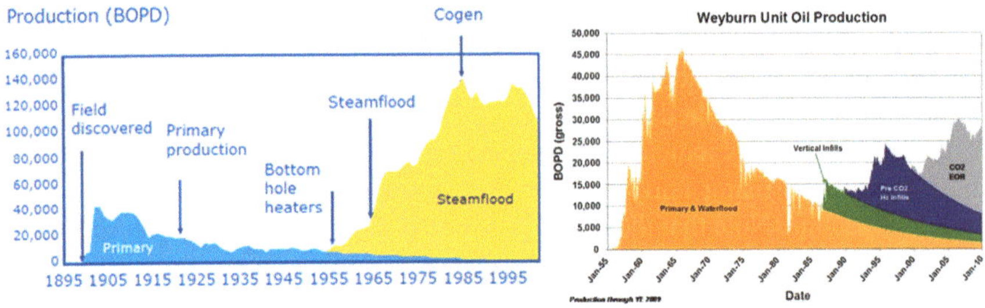

Fig. 2.2 Two fields with atypical production profiles.

Left Production of Kern River heavy oil field (13° API) in California, discovered in 1899. Long decline from primary production, then significant increase from the use of bottom hole heaters and subsequently steam injection. Today the field has many thousands of wells, achieving an average of ~15 bbl/day per well.

Source Chevron (from Google Images).

Right Production from the Weyburn field, Saskatchewan, discovered in 1955, showing large gains from CO_2 injection. (See Fig. 2.24, below, for IEA estimate of the total quantity of oil likely to result from use of this EOR technique.)

Source Cenovus Energy (from Wikipedia)

(but correctly, to show gains possible from EOR in specific fields) is in the IEA's *Resources into Reserves* (2013), p. 75. While such fields do exist, and are important to understand, they are far from common in number, and not significant in volume of oil compared to the global total.

(ii) Very large fields under the control of OPEC countries (and often the 'Seven Sisters' before that).

By contrast, the second class of 'atypical' fields is indeed fairly common, and because of the field sizes, very important. These are the larger OPEC fields, such as Aghajari above, where fields tend to have long flat profiles, partly due to their size alone, but mainly due, later, to OPEC quotas; and, earlier, to commercial restrictions on production to help limit global over-supply (Yergin 1991). In addition, production in such fields has often been punctuated in response to quota changes, or external events such as politics or war. The impact of these fields is discussed more fully in the sections below on regional and global peak.

2.1.3 Examining Field Decline, and Hence Field 'Ultimate' by a Linearised Decline Curve

Note that there is a useful way to analyse the decline in a field. This is to plot the field's production versus its cumulative production (as opposed to vs. date, as for example in Fig. 2.1). This approach is shown in Fig. 2.3 for the UK Forties field:

On such a plot exponential production decline becomes a straight line, and in practical examples extrapolation of this (roughly) straight line to the abscissa indicates, *ceteris paribus*, the likely maximum quantity of oil that the field will yield, i.e., the field's ultimately recoverable resource (URR). In Fig. 2.3 three such possible 'ultimates' are shown, all close together; with the gain in yield following sale of the field by the original owner (BP) being clearly visible.

Another example of this approach is given later as Fig. 2.16 for the Samotlor field; and similar informative plots are available in Hall and Ramírez-Pascualli (2013) for the following fields: Cantarell (Mexico); East Texas and Prudhoe Bay (USA), Brent and Statfjord (North Sea), Yibal (Oman) and Cusiana (Colombia).

2.1.4 Economic Constraints on Raising a Field's Production Post-Peak

The physical constraints listed above that drive field production decline lead in turn to economic constraints on how quickly the field's oil might be produced once its production peak is past. Many techniques exist to raise production from a field in decline, but these are usually costly and often only slow the rate of decline; see the

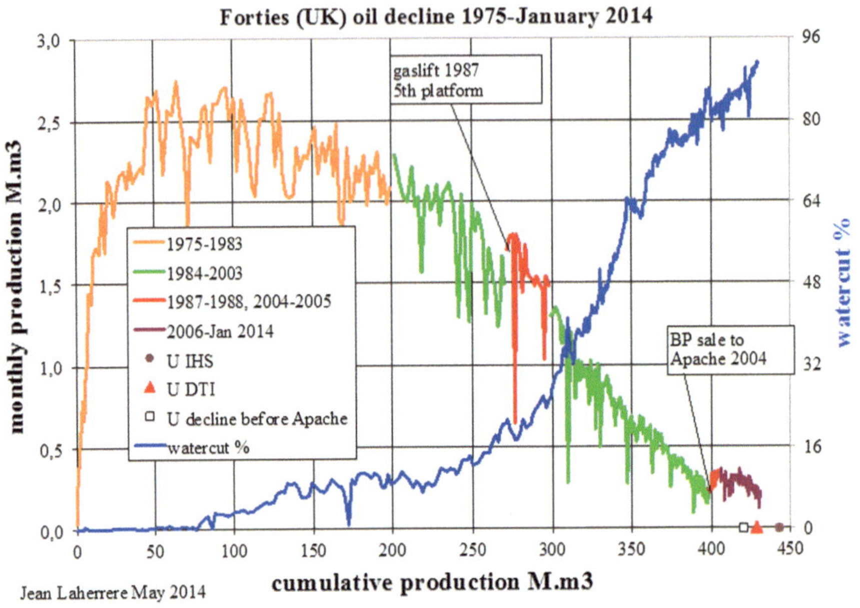

Fig. 2.3 Linearised plot of field decline: Example of the Forties field. Unlike a normal production plot of production versus date, this approach plots production (here as monthly data) versus cumulative production, and hence linearises the decline curve if the decline is exponential.

Notes

- Colours indicate production at different time periods.
- U IHS: URR estimated by IHS Energy.
- U DTI: URR estimated by the UK Dept. of Trade and Industry (data now from DECC).
- Right-hand scale shows increase in 'water cut' (water co-produced with oil as a result of water drive) over time, as a percentage of field total all-liquid (oil plus water) production.

Source J. Laherrère

general literature on reservoir engineering such as Muggeridge et al. (2014), or references such as Jakobsson et al. (2012) or Aleklett (2012). For specific fields see the examples of Magnus and Ula fields in Muggeridge et al.; or the Forties field shown here in Figs. 2.1 and 2.3. Operators of such fields optimising net present value therefore often find that even at high oil prices only relatively little extra oil (compared to the field total) can be produced profitably. UK production post-1999, where the total of production from all fields fell steeply despite the real-terms oil price rising over five-fold, from under \$20/bbl to over \$100/bbl, provides a good example of this (Fig. 2.10).

Note however that the quantity of 'extra oil' that can be produced from fields as the technology advances and if the oil price rises is important and needs to be understood, even if it perhaps does not affect the dates of peak by much. This is discussed in Annex 4, in the section on 'Reserves growth'.

2.2 Conventional Oil Production in a *Region* (i.e., a Group of Fields)

Now we turn from peak in an *individual field* to considering the peak of conventional oil production in a *region* containing a group of individual fields. Here two new factors enter the discussion: the field size distribution in the region, and the amount of conventional oil in the region that has not yet been discovered. Field size distribution is critical because the volume distribution of *conventional* oil in a basin is usually very asymmetric, with most of the oil being in a relatively small number of large fields. Such fields tend to be more easily found than smaller ones, and also brought on-stream earlier.

2.2.1 A Simple Model of Oil Discovery and Production in a Region

To understand the mechanisms that drive peak in a region consider a very simple model that reflects the trend of oil discovery and production that is typical of most regions. This is illustrated in Fig. 2.4. The grey bars indicate the assumed field size distribution in the region, with the largest field containing 100 units of oil, the next 90 % of this, and so on. The model is also simple in that it assumes that exploration effort allows only one field to be found a year, and that the fields are found in size

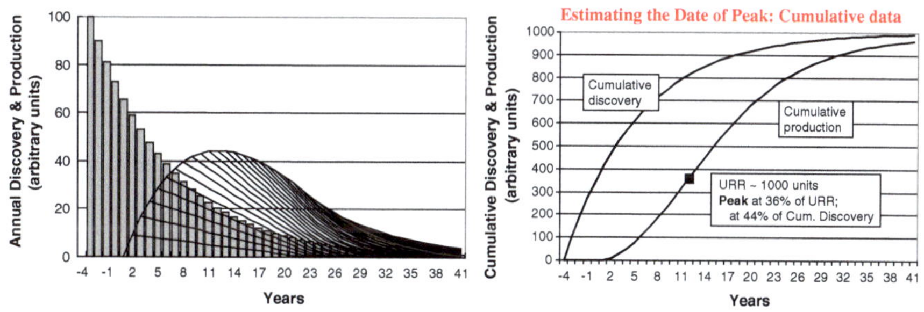

Fig. 2.4 A simple model of conventional oil discovery and production in a region.
Left Shows the field size distribution and discovery sequence (*grey bars*), and each field's subsequent production (*triangles*), where each field is assumed to take 4 years from discovery to production. The plot is to-scale such that for example the volume of oil shown as discovered for Field 1 (leftmost grey bar, 100 units) is the same as indicated for Field 1 production (the lowermost production triangle, which starts in Year 1, reaches 9.09 units/yr. in Year 2, and falls to zero by Year 23).
Right The same data for discovery and production, but on a cumulative basis. The *resource-limited* peak in production (at Year 12) is denoted by the small *solid square*

order. That is, as the Figure shows, the largest field is found in Year −3, the next largest in Year −2 and so on.

The model then assumes that it takes four years for a field to get into production. Thus production of the largest field starts in Year 1, the next largest in Year 2, and so on.

Finally the production profile assumed for each field is also simple: production rises over the first year, reaches a peak in the second year, and then declines linearly thereafter, with a total life of 22 years. The plot is to-scale, so that the total area (i.e., volume of oil) under the production curve for the each field is the same as that shown by its corresponding discovery bar.

On these assumptions, the total production curve for the region emerges. As can be seen, in this case the region reaches a peak when about one-third of the region's total oil has been produced. Despite the simplicity of this model, this general 'whale-back' shape for a region's production curve of conventional oil is surprisingly valid. It roughly matches what has occurred in the majority of the sixty or so oil producing countries that are past their conventional oil peak, provided they avoided major disruptions in production, and also the profile given by Hubbert in his early publications and in interviews on film.[3] Examples of 'real-world' production graphs for a number of countries split by field are given in Figs. 2.7, 2.10, 2.13, 2.18 and 2.19 below; while production graphs (but not split by field) for virtually all oil producing countries are in variously Hallock et al. (2014), Campbell (2013), or on the Globalshift Ltd. website (www.globalshift.co.uk).

The left graph of Fig. 2.4 is very telling. It shows the main drivers of the peak of conventional oil production in a region, and explains why this *production typically reaches a maximum when something approaching only half or less* of the region's total conventional oil has been produced. Importantly, as the graph shows, the peak in a region is driven by:

- the asymmetry in oil location: most of the oil is in a small number of large fields;[25]
- the fact that these large fields tend to be discovered first;
- production in individual fields declines;
- more fields are being discovered, but they contain much less oil, and hence production in the region peaks.

Note that as long as many new fields are being discovered that contain significant quantities of oil, then the added production of these fields can offset the decline from earlier fields. The *resource-limited* peak of conventional oil production in a region thus occurs only when discovery in the region is well into decline. It is for this reason that knowing the true discovery history of a region is key to understanding the region's potential for future production.

2.2.2 Realities Behind This Simple Model

Of course the above model is over-simple, and in reality basins—and even more so larger regions—can be complex; and new plays can open up as the geological knowledge and technology advances. In such cases the discovery trend for a region can display 'multiple asymptotes', and it takes geological and engineering knowledge to judge when overall discovery in the region is drawing to a close, as in the Gulf of Mexico for example.

Moreover, the simple assumptions of the model reflect an interweaving in the real world of geology, engineering and economics. The rate that fields are discovered in a region, and then brought on-stream, is affected by the geology of how easy the big fields are to find versus the smaller later ones; how fast the geological and engineering knowledge of fields builds up; and the economics that determines the initial search effort, the rate that fields are brought into production, and their production histories. It is possible, for example, for a surge of small fields to be brought on-stream rapidly, as happened with the UK in 1998 when the oil price was low and companies sought to maintain revenues by production increases. But overall the model is reasonable in capturing the essence of oil discovery and the resulting production.

Note that the key feature of any realistic oil 'discovery-and-production' model for conventional oil is that the volume of oil discovered in fields in a basin typically gets less over time. In the simple model of Fig. 2.4 this occurs because discovery is restricted to one field per year, and since fields are discovered in size order the discovery volume per year automatically falls. In another example, Bardi and Lavacchi (2009) propose a simple two-equation model that relates oil production to capital expended, where an increasing quantity of capital is required over time for a given quantity of production. With suitable parameters the model replicates the Hubbert curve; but it can also be applied to other resources, and can capture the falling energy-return ratios of many resource extraction histories.

The Global Energy Systems Group at the University of Uppsala has modelled the size distribution of fields in a region, their discovery rate, and time-to-production on the basis of appropriate probability distributions to generate the expected production profile of the region, and is currently improving these models. Many other oil production models have also been proposed, see for example the review by Brandt (2009). The majority of these models tend to generate a roughly Hubbert or 'whaleback' regional production curve over time.

Finally, Hall (private communication) postulates that the fall-off over time in the volume of oil discovered in a region is a *direct consequence* of the increasing amount of energy required to search for, and to bring on-stream, the progressively smaller fields in the region.

2.2.3 Summarising Findings from This Simple Model

Summarising the findings from this simple model of oil discovery and production of *conventional oil* (oil in fields) in region we can say:

(a) The resource-limited peak of conventional oil production in a region is caused by *adding the output of successive fields, where the later fields are generally smaller than the earlier.* This reflects the fact that the size distribution of fields in most areas is very skewed, with most of the oil being held in a relatively small number of large fields that tend to get found first.

(b) The peak occurs once *discovery has declined significantly*; and indicates the point at which reduced output from the early fields is no longer compensated by increased production from the later. The typical shape of the regional production curve is driven by the profile of decline in individual fields, primarily from field pressure loss, reduction in effective oil column, and possibly from increasing drive fluid bypass. In the case of this simple model, the peak of discovery is 16 years before the production peak.

(c) If a region sees significant separate phases of discovery, such as on-shore followed by offshore, then production may also show *a number of resource-limited peaks*, each reflecting a different discovery phase.

(d) Finally, it is important to recognise that the production peak is *counter-intuitive*. This is because the peak occurs, as Fig. 2.4 shows, when:

- production has been trending steadily upward;
- the remaining reserves are large (and generally low-cost, as know-how and infrastructure are in place for their development);
- discovery is continuing;
- technology is improving, and hence recovery factors likely to be increasing.

2.3 Predicting the Peak of Conventional Oil Production in a Region by Combining the 'Peak at Mid-Point' Rule with the 2P Discovery Trend

Now we turn to the question of what the simple model of Fig. 2.4 can tell us about *how to predict* the production peak in a region.

It is a truism to state that oil cannot be produced unless it has been discovered. To examine the *discovery* trend in a region, and in turn to understand the scope for future production, the data in the model of Fig. 2.4 (left) are best presented as a cumulative plot, Fig. 2.4 (right).

As this shows, in this simple model about 50 % of the final discovery has already occurred by the time production of the first field starts in Year 1. And by the time production peaks in Year 12 the discovery curve has turned well towards its asymptote. In the real world also, and critical for prediction purposes, a region's discovery asymptote is usually clear well before the production peak has occurred.

The next step is to apply the approximate 'peak at mid-point' rule. This says that the '*resource-limited*' production peak of oil production in a region typically occurs when roughly half of the region's URR has been produced; and where the region's URR is given by the asymptote of discovery.

This key idea is worth restating: To be able to predict when a region's '*resource-limited*' peak of conventional oil production is likely to occur the region's *discovery trend* can be combined with the 'peak at mid-point' rule. (As already mentioned, the discovery trend needs to be based on the oil-industry 2P data, not the misleading public-domain 1P data.)

Figures 2.6, 2.9, 2.12, 2.15 and 2.17, give examples of this approach for a range of countries. These Figures give the oil-industry 2P oil discovery data for these countries, and also their oil production, as both annual and cumulative plots. These plots allow the date of peak conventional oil production to be estimated, whether this peak is past or in the future.

2.3.1 The IHS Energy 'PEPS' Data

A number of the Figures used in the discussion below present data from the IHS Energy 'PEPS' dataset, and are provided with the company's permission. In understanding 'peak oil', these data (or the equivalent from other industry 2P data sources) are key, and it is important to understand how they are generated, and hence their reliability.

A consultancy to assemble oil field data was set up by Harry Wassall in 1956, originally based in Havana. This became Petroconsultants S.A. when the headquarters moved to Geneva in 1968, and it was later bought out by IHS Energy in 1996. The company's database collects, *inter alia*, data on oil and gas wells, and fields, from around the world, and aims to give global coverage. The data are 'scout', in that they are assembled by company employees scouting for information from a wide variety of sources. In the early days (and to some extent still today) this was done mostly by personal contact within the oil companies; and where often the latter, not allowed legally to discuss data with rival organisations, were happy to share data with the consultancy in exchange for access to data which the other companies were willing to supply.

When the 'Reading Oil Group' first encountered the issue of 'peak oil' in about 1995, much of our effort went into understanding the data that the various proponents for and against peak oil were using. It became clear that while other commercial oil and gas field datasets existed at that time (and more now), that of Petroconsultants' was generally seen by the oil industry as preeminent, especially in its degree of international coverage. These data were the basis of the Petroconsultants' studies that led to the Campbell and Laherrère 1998 *End of Cheap*

Oil article, and were used in the USGS Year-2000 Assessment, and probably in subsequent assessments. Though the data are widely purchased by oil companies, the full dataset has been seen as too expensive to purchase by some of the national and international energy agencies.

Over several years, our 'Reading Oil Group' had useful conversations with Dr. George Leckie of Petroconsultants, who at that time was responsible for entering many of the estimates of oil and gas field size into the database. These estimates were seen as specifying the most likely amount of oil or gas a field would produce over its lifetime, in light of both currently committed infrastructure and technology and what might reasonably be assumed in future. Such estimates were taken to reflect the nominal 'mean', or proved-plus-probable (2P), values for each field, and hence contrasted sharply with the proved-only (1P) data that oil companies were required to report publically under SEC rules.

Since the production by the individual fields was also recorded, cumulative production for a field can be deducted from the estimate of field size to give the field's remaining 2P reserves. In addition, for each field the company registers the field's date of discovery. Moreover, because the data are notionally statistically mean values, data for individual fields can be added arithmetically within the dataset to yield basin, country and global totals.

Then, since the aim was (and is) to capture 2P information on all fields globally (except for non-frontier US and Canada, where the data are only 1P), a picture is generated of how much oil or gas has been discovered in a region at a given date, and hence determine the region's corresponding trend of 'true' (2P) oil or gas discovery over time. It is these 2P discovery data, and the corresponding production data, that are given by country in the Figures below.

One problem however arises with these data, at least as far as analysis is concerned. The company (almost certainly like other data providers) was generally requested by customers for the best *current* estimates of the size of fields. This meant that if a revised estimate for the original size of a field became available, the database was simply updated with this new number. That is, for example, data in the database for the year-2000 (as given in the Figures below) reflects year-2000 knowledge of the size of fields, and not the estimates made at the dates the individual fields were discovered. Such data are said to be 'backdated'; and where, as a result, information on how the estimate of a field's size has changed over the years (its 'reserves growth') has become lost; unless earlier versions of the database are accessed, which can sometimes be done. Recently, IHS Energy has begun recording the change in field size by date in some fields; but a 'phone call to the company a short while back confirmed that such 'reserves growth' data are not yet available on a global basis.

A second aspect of the data that analysts need to be aware of is that originally the data were largely (or entirely) only for oil in fields (i.e., for conventional oil). Today, with the growth of production of the non-conventional oils, the database contains information on the oil volume expected from *specific projects* of non-conventional oil production. As a result, for example for the Canadian tar sands, estimates of the volume to be produced from announced projects are available in the database, but not (and quite correctly so) estimates of the total amount of tar sands oil potentially recoverable, despite the fact that all this oil might be classed as 'discovered', in the sense that its general location is already known.

The extensive exploration and production data (on well, field, project, seismic and related statistics) are held by IHS Energy in their Edin database, and this is updated on a continuous basis. The data in the company's 'PEPS' database, such as those used below, are extracted from Edin at a by-country level on an annual basis.

Incidentally, readers may be curious why 'PEPS' data only to the year 2000 are given in the Figures below. The answer is simple: release of these data needs permission from the company, and this has only been given for the Figures provided here. However, as these show, by the year 2000 the 2P discovery trends in these countries (and, indeed, for virtually all oil-producing countries globally) were past their respective 'discovery-trend' inflection points; signifying that the rates of discovery in these countries were in decline by that date, such that in these (and in nearly all) countries, provision of later data would not change the overall trends of decline in discovery by much.

2.3.2 *Oil Discovery and Production in Germany*

With this information in hand, we can now look first at Germany, as this is an example of a country where its oil production peak is long past. We cover Germany in some detail as it illustrates most of the principles that underlie the peak of conventional oil production in a region. We start with Fig. 2.5 (left), which shows German 'liquids' production (here, meaning crude oil plus NGLs) from 1900 to the mid-1960s.

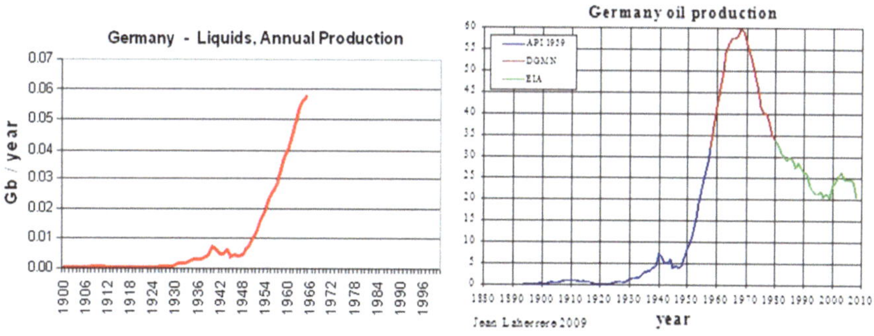

Fig. 2.5 Germany: Annual production of 'liquids' (i.e., crude oil plus NGLs).
Left chart Production 1900 to the mid-1960s.
Source IHS energy, with permission.
Right chart production 1880–2008.
Source Jean Laherrère (colours indicate data sources, as listed)

As can be seen in the left chart of Fig. 2.5, there was a small peak in German fossil oil production during World War 2, but the main feature was the rapid growth of production in the subsequent years. At the end-date of this chart probably most analysts would be tempted to extrapolate this production trend on upwards. Not only was this production trend looking robust, but at that date Germany had plenty of oil left in proved reserves; discovery was known to be continuing (indeed she had not yet discovered her largest field); and technology—particularly tertiary recovery—was being introduced that was significantly raising recovery factors in existing fields. Every reason, one might think, to expect production to keep rising.

Needless to say, such an expectation would have been very unwise, as shown in Fig. 2.5 (right), as production immediately peaked. But a key question is then raised: was this peak the country's conventional oil *resource-limited* peak, or was it due to other causes? Maybe the government limited oil production because of environmental concerns over drilling; or pro-rationed output due to regional or global over-supply as happened in the US; or tried to raise the oil price in the manner of an OPEC-style quota; or was oil production in Germany simply too expensive, and the country found it easier to obtain her oil from cheaper sources outside?

This reflects a serious problem with observing only oil *production* data: the data may show a production peak but is this peak *resource-limited*, such that production must continue to decline; or are there other factors at work, and production can go back up? The answer lies in examining the oil *discovery* data, and this is done in Fig. 2.6.

In Fig. 2.6 (left), the blue bars show the amount of oil discovered (as measured by 2P data) in Germany for each of the years shown, while the red line gives the corresponding amounts of oil produced in these years. As the blue bars show, some oil was discovered in the country before 1900, but the bulk of the country's oil was discovered in the 1940s and 1950s, where this was due to the introduction of the use of seismic. The large Mittelplate field (included in the large blue bar for discoveries made in 1980) was a late find, made in Germany's rather small offshore area that then became open for exploration. As can be seen in Fig. 2.6 (right), Germany's cumulative on-shore trend of 2P discovery had flattened out since about 1960; and when combined with the oil offshore, since 1980.

As the Figure makes clear, the production peak seen in Fig. 2.6 (left) was indeed *resource-limited*, at least in the sense of being driven directly by the amount of oil that had been discovered up to the year 2000. Had more oil been discovered, this production peak could have been higher or later; if less discovered, then lower or sooner. Simple as that: discovery controls production; once you know discovery, you know the limit to corresponding future production.

In terms of comparing Germany's output to the simple model of Fig. 2.4, we can ask: what was happening at field level? This is shown in Fig. 2.7.

On the basis of Fig. 2.7, clearly the model of Fig. 2.4 is an over-simplification for these larger mostly on-shore fields. The numbers of years before field peak is reached is often quite long, and hence not almost immediately as with the triangular production profiles of the simple model. But the general mechanism of peak—of a

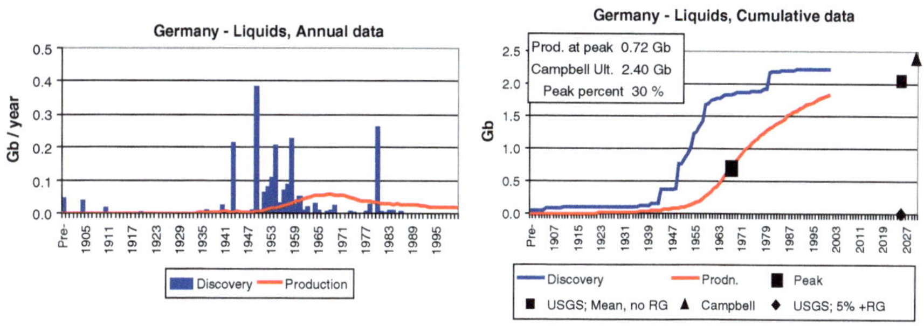

Fig. 2.6 Germany: Oil-industry '2P' data on oil discovery, and production; 1900–2000. Data are for the IHS Energy definition of 'liquids'; here meaning crude oil plus NGLs.

Left chart

- *Blue vertical bars* Annual 2P oil discoveries. Data are year-2000 backdated data; i.e., reflecting information available at the year 2000. Volume discovered prior to 1900 is indicated as 'Pre'. Large late find is the Mittelplate offshore field in 1980. Peak of discovery in ~1950.
- *Red line* Annual production. Peak of production was in 1967.
  Note In this chart (and in the equivalent charts below) the width of the 'discovery' bars (*blue*) is set to one year; so that the area of these bars and the corresponding final area under the production curve (*red*) will be equal if all the oil discovered is eventually produced.

Right chart Cumulative plot of the same data.

- *Blue line* cumulative backdated 2P discovery.
- *Red line* cumulative production.
  This plot indicates that the production peak (indicated by the solid medium-sized square) is indeed *resource-limited* based on the quantity of *oil discovered to the year 2000*; with the peak occurring at ~30 % of a URR that would seem reasonable from extrapolation of 2P discovery. (Also shown, URR estimates from the USGS and Campbell, see text.)

Source IHS energy 'PEPS' year-2000 dataset, with permission

number of the larger fields on-stream early, plus the later decline in these fields—is clear. Also note in this Figure the decline that would have occurred in Germany's oil production had Mittelplate had not come to the rescue.

2.3.3 *Could There Be a Later Peak? I.e.: Is the 2P Discovery Trend a Reliable Indicator of URR?*

Now we address an important question: Is the 2P discovery trend reliable, in the sense that extrapolation of this trend gives a useful indication of a region's conventional oil 'ultimate' (URR)? This depends on there not being large new discoveries of conventional oil waiting over the horizon. The only way to answer this question is to ask the petroleum geologists what more conventional oil is likely to be found in the region.

In Fig. 2.6 (right) two geologically-based estimates of Germany's conventional oil 'ultimate' are given: that of the USGS in its year-2000 assessment, where the

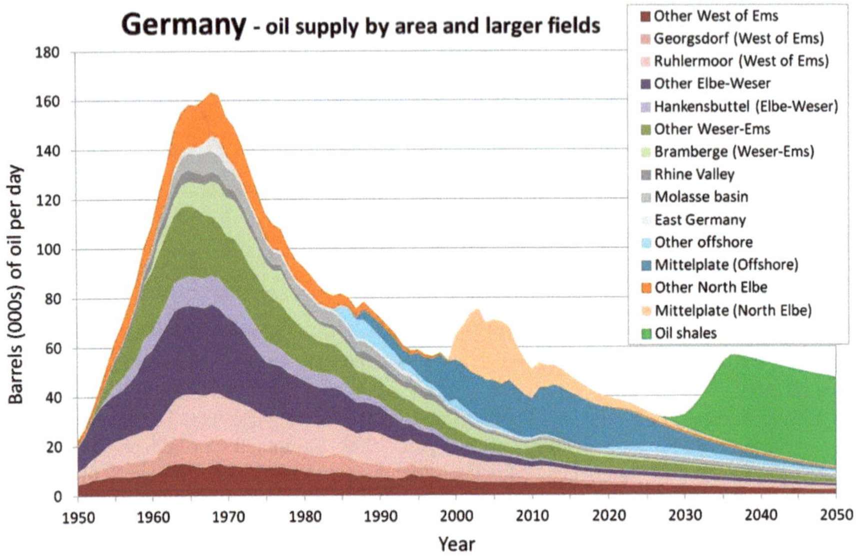

Fig. 2.7 Germany: Oil production by largest fields in regions, and by other fields in these regions. Actuals: 1950 to ~2013; Forecast: ~2013–2050.

Notes Largest fields are Georgsdorf, Ruhlermoor, Hankensbuttel, Bramberge, and the Mittelplate primarily offshore field. Indicative production from oil shales shown coming on-stream later.

Source Globalshift Ltd

'mean, no-reserves-growth' estimate is shown by a small square at 2025; and that of Colin Campbell, an estimate at about the same date, shown by the triangle at 2030.

As can be seen both estimates are in general agreement with each other, and with a reasonable extrapolation by eye of the cumulative 2P discovery trend up to the year 2000. Thus on the basis of these two estimates, we can say that in the case of Germany extrapolation of the 2P discovery trend does indeed give a reasonable estimate of URR; and hence, using the 'peak at mid-point' rule, in turn allow an estimate to be made of the region's date of peak conventional oil production. (Note that as the USGS data sum only basins evaluated, their URR estimate may have excluded Germany's offshore.)

In order to underline this relationship between extrapolated discovery and URR, in Fig. 2.8 three further geologically-based estimates of Germany's conventional oil 'ultimate' are plotted.

The URR estimates shown in Fig. 2.8 all apply only to conventional oil, and are:

- BGR's 1997 assessment of estimated ultimate recovery ('EUR'): 2.3 Gb;
- USGS' year-2000 median assessment on a 'non-grown' basis, incl. NGLs: 2.14 Gb (may exclude N. Sea fields, as mentioned above);
- Campbell/University of Uppsala end-2004 model, excl. NGLs: 2.75 Gb. [Note that this URR is thus up from the earlier estimate of Fig. 2.6 (right) of 2.4 Gb;

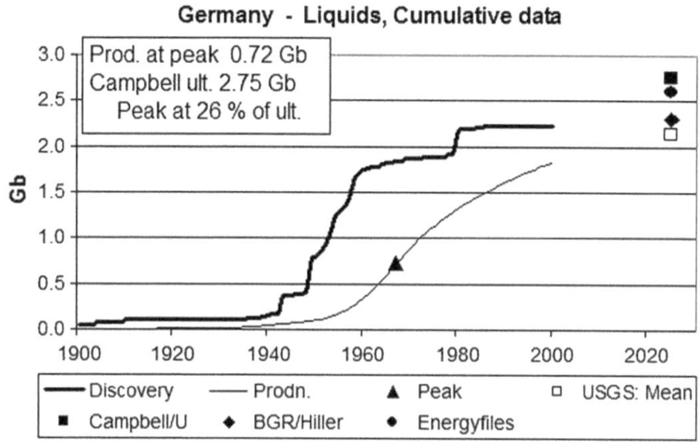

Fig. 2.8 Germany: Cumulative oil liquids discovery (2P data), and production, 1900–2000; and estimates of that country's conventional oil 'ultimate' (URR).

Sources Discovery and Production: IHS Energy, with permission; 'Ultimates': BGR (1997), USGS (2000), Campbell/Uppsala, (2005), Energyfiles (2005).

Notes

- The date of the production peak is marked with a triangle.
- URR estimates made around the year 2000 are shown against the year 2025, as notionally this was the end-year that applied to the USGS estimate. In practice all four 'ultimates' refer to much later dates. Three of the groups recognise that future extraction technology and policies are unknown, so specifically caution that their URR estimates should not be seen as definitive estimates of 'true' ultimates (i.e. original endowments of recoverable conventional oil when extraction terminates). Instead the estimates refer to quantities of oil considered recoverable over reasonably long time spans. The USGS said they evaluate oil that will be available for discovery by 2025 (though there has been ambiguity about this date). The Campbell/Uppsala model no longer lists ultimate, but 'Regular conventional' oil production to 2075'; ('Regular conventional' oil here excludes polar, deepwater, very heavy oils and NGLs). Energyfiles' URR quantifies oil that will have been produced by 2145. The BGR is the only organisation that uses the label 'estimated ultimate recovery', but probably would apply the same caveat as the others if asked

and compares to the later estimate of Campbell (2013) for Germany's 'Regular conventional' oil of 2.50 Gb (vs. cumulative production to 2010 of this oil of 1.96 Gb).]

- Energyfiles end-2004 assessment: 2.6 Gb.

As the Figure shows, these 'ultimates' are in rough agreement with each other, and with the apparent asymptote of the 2P discovery curve. The geologists were therefore fairly certain that no significant new quantities of conventional oil would be found in Germany, where this reflected both geological knowledge and over a hundred years' of discovery effort and technological progress.

Note also that like other regions of the world, Germany, despite having applied enhanced oil recovery (EOR) techniques since 1985, still had (and has) a considerable amount of oil judged currently unrecoverable in existing fields. However, barring some extraordinary new recovery technique, Germany by the year 2000 was judged close to the end of her conventional oil: at ~2.0 Gb Germany's total production by that date having consumed about 80 % of her estimated recoverable original endowment of this oil.

In comparison to the 'mid-point' rule, Germany's production peak occurred at only 26 % (not 50 %) of the Campbell/Uppsala ultimate of 2.75 Gb. But the peak reflected only the *on-shore* fields being produced at that date, so the peak occurred at about 35 % of the apparent on-shore discovery asymptote, of about 2 Gb.

In summary, Germany is an example of a region where there would seem to be little scope remaining for new discoveries; where total recoverable conventional oil is now nearly fully depleted; and where its conventional oil production peak in 1967 was indeed *resource-limited.*

We have examined the case of Germany rather fully in order to set out the principles of using 2P discovery data to investigate a region's production peak. Next we look—rather more briefly—at the 2P discovery and production data for a number of other countries.

2.3.4 Oil Discovery and Production in the UK

Figure 2.9 applies to the UK, and shows the oil-industry 2P discovery data, and also production. The left chart gives annual data to 2004, and the right chart the equivalent data (albeit from a different industry source) to the year 2000 in cumulative format.

As can be seen, the UK has had two oil production peaks, with the trough between these being caused mainly by safety work carried out on all fields following the Piper-Alpha disaster in 1988. Lesser factors for this production trough include the 2-year work-over on Brent due to high gas production; the fall in oil prices post-1984; anticipated changes in petroleum revenue tax that may have delayed the start-up of new fields; and—as Laherrère notes—the secondary peak in discovery in the late 1980s, as indicated in Fig. 2.9 (left). Without this 'trough', production most likely would have risen to a peak in the early 1990s. As it was, the UK's conventional oil resource-limited peak occurred in 1999, at 59 % of Campbell's estimate of URR.

To compare to the simple model of Fig. 2.4, Fig. 2.10 gives the corresponding plot of UK production broken down by field.

As the Figure shows, except for the 'trough', this plot of a region of mainly offshore fields is more like the simple model of Fig. 2.4 than the corresponding plot for Germany, which is primarily of onshore fields. This is not surprising, as the simple model was originally devised from examination of UK data.

Although discovery is not broken out by field in Fig. 2.9 (left), the pattern of UK discovery roughly matches that of Fig. 2.4; once a small initial field had been discovered in 1969, nearly all the very large fields were discovered fairly rapidly thereafter. As can be seen, the explanation is supported that peak is caused by a

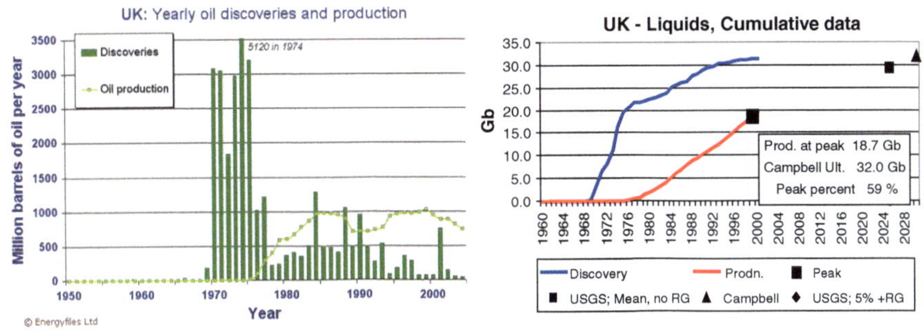

Fig. 2.9 UK: Oil-industry '2P' data on oil discovery, and production.

Left chart Oil-industry '2P' data on oil discovery, 1950–2004.

- *Green vertical bars* Annual 2P oil discoveries (data are year-2004 backdated; i.e., reflect information available at 2004). Peak of discovery in early 1970s. The medium-sized late find is Buzzard, discovered in 2001.
- *Green dotted line* Annual production. Peak of production in 1999.

Source Energyfiles Ltd. with permission.

Right chart Cumulative plot of the similar data, 1960–2000.

Data are for IHS Energy definition of 'liquids'; here meaning crude oil plus NGLs.

- *Blue line* Cumulative backdated 2P discovery.
- *Red line* Cumulative production.

This plot indicates that the production peak (indicated by the solid medium-sized square) is indeed *resource-limited* based on the quantity of *oil discovered to the year 2000*; with the peak occurring at 59 % of a URR that would seem reasonable from extrapolation of 2P discovery. (Also shown, URR estimates from the USGS and Campbell, see text.)

Source IHS Energy 'PEPS' year-2000 dataset, with permission

region's large fields mostly getting into production first and then declining. In addition, by comparing the volume discovered with the volume produced, and using the 'peak at mid-point' rule, it was clear that the 1984 peak was not resource-limited, while the 1999 peak would appear to be.

But now we have to return to the important question raised above for Germany. How are we sure that the UK's 1999 peak was indeed resource-limited? This is clearly the case if based on the oil already discovered by that date; but how do we know that the UK does not have has big new plays of conventional oil waiting in the wings that will yield enough oil to surpass the 1999 peak? As already mentioned, this situation can occur where the historical discovery data (the 'creaming curve' vs. time) indicates an apparent asymptote, but where this asymptote increases as a major new play enters the scene.

The answer, as already mentioned in connection with Germany, is that knowledge of peak cannot be based *solely* on discovery data, it must also include geological appraisal. It is recognised that the latter will always be judgement, and that the chance of future large finds cannot be known with certainty. But a great deal of geological knowledge now exists for much of the world's likely oil plays, and as explained later, *globally* the discovery of conventional oil in new fields has been

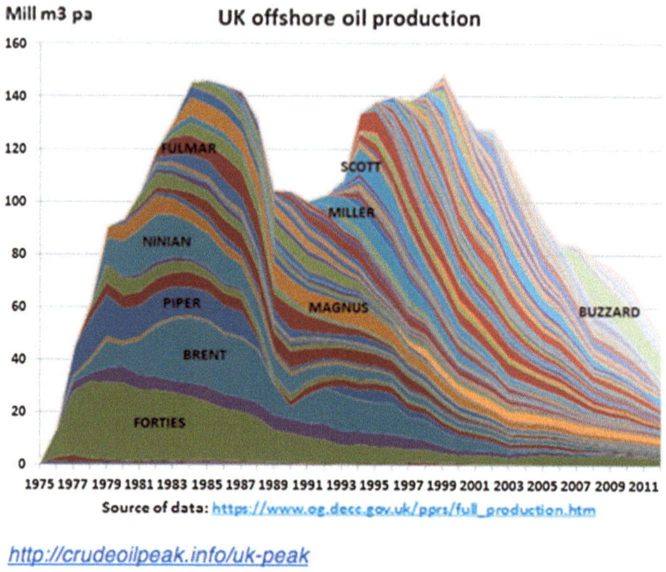

Fig. 2.10 UK: Production by field, 1975–2011.

Notes

- Compare Forties output to the decline plot of Fig. 2.3.
- Note loss of output, and later restoration, of Piper output; and significant reductions and recoveries in other large fields due to associated safety work; and also the Brent work-over.
- Numerous small late fields cannot compensate for the decline of the early major fields.

Source M Mushalik (see website cited on the Figure)

falling for about half a century, so the scope for surprises in terms of big new discoveries is now judged generally as rather small.

Note however that in the UK's case, as elsewhere, even for conventional oil there are still significant future *potential* sources of oil. Some experts suggest that there remain quite large quantities of UK oil undiscovered in subtle stratigraphic traps; there is certainly still potential in the deeper Atlantic; and there is known to be a large amount of oil currently in-place in existing fields deemed unrecoverable with today's technology and oil price.

But geological and reservoir knowledge says it is virtually certain that none of this UK oil, where it exists, can be developed rapidly enough to push production back up past the 1999 peak. The subtle traps, if they hold significant amounts of oil, will need highly calibrated seismic to find, so will not be found rapidly; the deeper Atlantic will offer surprises but is not thought especially prospective due to poor source rock and traps; while the many routes to improved recovery (EOR) in existing fields have already seen much trial and analysis. Overall, combining the UK's 2P discovery data with geological knowledge indicates that the country's *conventional* oil peak in 1999 was indeed *resource-limited.*

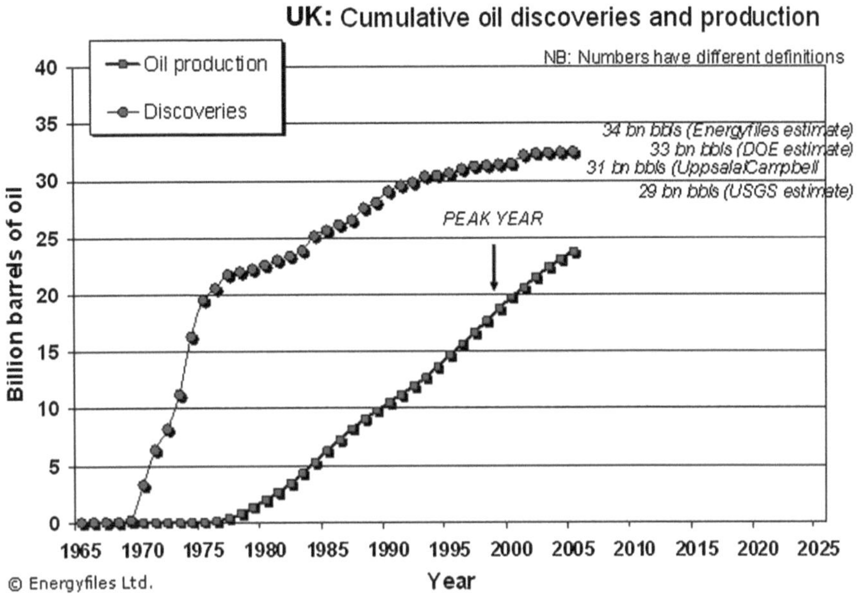

Fig. 2.11 UK: A second plot of '2P' oil discovery, and production, displayed on a cumulative basis. Also shown are four estimates for the UK's conventional oil 'ultimate'. The UK Department of Energy's estimate ('DOE') is from 1974; the others are more recent: USGS in year 2000; and Campbell/University of Uppsala and Energyfiles Ltd. estimates made about 2005. The USGS year-2000 and Campbell/Uppsala estimates exclude NGLs (these add ~4.5 Gb); the USGS estimate also excludes UK West of Shetlands basins. [*Note* Campbell (2013) estimates 'Regular conventional' URR as 32 Gb; and cumulative production to 2010 as 24.7 Gb.]

Source Energyfiles Ltd.

Figure 2.9 (right) brought out this point by giving two estimates of the UK's ultimately recoverable conventional oil resource (URR); that of the USGS Assessment of year-2000; mean value and without allowance for reserves growth, and Campbell's of about the same date, both ex-NGLs. As with the case of Germany, these estimates are close to each other, and also broadly in agreement with the value that might be expected from extrapolating by eye the backdated industry 2P discovery data. Three further estimates of the UK's conventional oil URR are given in Fig. 2.11. The earliest is a UK government DoE 'Brown Book' estimate made back in 1974 (see Annex 2); and the more recent are those from University of Uppsala/Campbell and Energyfiles Ltd, both made around 2004. As the Figure shows, these 'ultimates' are again in close agreement with each other, and with the asymptote of the '2P' discovery creaming curve.

2.3.5 *Expecting (and not Expecting!) the UK Production Peak*

Given that nearly all of the UK's large fields, and over half of the UK's total oil, had been discovered by the time offshore production started in 1975, it is not surprising that realistic estimates of the UK's conventional oil ultimate were available from an early date. These included the UK Department of Energy's 1974 estimate of 4500 million tonnes (33 Gb) shown in Fig. 2.11; and see also the additional early URR estimates given in Annex 2.

Then using the 'mid-point' rule—well known and well understood at the time—it was easy to predict that UK production would peak at, or probably a bit before, the point where about half of this (i.e., 16.5 Gb) had been produced. Looking at Fig. 2.11, this meant around the mid-1990s if the slowdown due to the 'production trough' is ignored. It was this understanding of the likely date of peak—well known within the industry—that allowed a 1976 UK research study for the government to note that the date of the world oil peak (at "about [the year] 2000") would not be far behind that of the UK peak (UK Department of Energy 1976, p 12).

However, somehow this information on the UK peak got lost. In about 1997 and 1998 our small *ad hoc* Oil Research Group at the University of Reading tried several times to warn the UK's Dept. of Trade and Industry (DTI) of the coming *global* peak of conventional oil production, where our line of argument was simple: 'You understand the mechanism behind the coming UK peak, and you know that this is close; the world peak works in rather the same way, and the 2P discovery data show that is fairly close also'.

Unfortunately, this argument fell completely flat. The concept of 'mid-point peak' had been forgotten (and not just in the UK), and a deep myth had developed instead based on the evolution of *proved* reserves. In the UK, for example, UK *proved* reserves had held steady since about 1980 at between 4 and 5 Gb, despite annual production being nearly 1 Gb/y for most of this period (Annex 2). As a result, this nearly two decades of there apparently being only roughly '5 years' supply' of UK oil remaining (plus corresponding data for other countries) had fooled nearly all analysts (including many within the oil industry, and most within the UK government and also at the IEA) into thinking that this 'replacement of reserves' was primarily due to improvements in technology; with horizontal drilling and 4-D seismic being widely cited. Our arguments about the proximity of the global oil peak were therefore seen as baseless, because the DTI were convinced that the UK peak itself was still many years away; and that afterwards production would decline only gradually anyway, because of future gains in technology. (For a fuller discussion of these meetings with the UK's DTI and with other government bodies, see Chap. 4 of Campbell (Ed.) 2011.)

Nevertheless, despite the DTI's scepticism, the UK peak was indeed close. Today we can give the UK as an example of a country where, for *conventional oil*, only modest scope remains for new discoveries, where the likely total recoverable quantity of this oil is judged well over half depleted, and where the country's resource-limited oil production peak of this oil is past.

2.3.6 *Oil Discovery and Production in Norway*

Next we look at comparable graphs for Norway. These are given in Figs. 2.12, 2.13 and 2.14, and indicate similar findings to those for the UK and Germany.

As Fig. 2.12 shows, Norway's discovery has come in 'lumps'; while her production rose to 1997 and then fell slightly. Production then peaked in 2001 (as shown by the medium-sized square in the cumulative chart). The latter plot indicates that this production peak was indeed resource-limited, if based on the quantity of oil discovered to the year 2000; with the peak occurring at 47 % of a URR that would seem reasonable from extrapolation of 2P discovery.

Note that Norway, despite her closer involvement with the oil industry via her government share of StatOil, had, like the UK, serious issues in the early years with the misleading evolution of her proved reserves, and only later got the proper presentation of the reserves data sorted out.

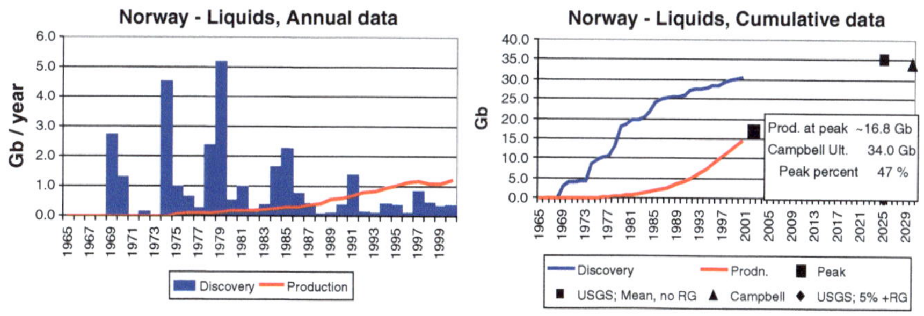

Fig. 2.12 Norway: Oil-industry '2P' data on oil discovery, and production; 1965–2000. Data are for the IHS Energy definition of 'liquids'; here meaning crude oil plus NGLs.

Left chart:

- *Blue bars* Annual 2P oil discoveries (data are year-2000 backdated; i.e., reflect information available at 2000). Peak of discovery in ~1980.
- *Red line* Annual production. (Peak of production was just after the end of this plot, at 1.25 Gb/y in 2001.)

Right chart Cumulative plot of the same data.

- *Blue line* Cumulative backdated 2P discovery.
- *Red line* Cumulative production.

Also shown, URR estimates (made about the year 2000) from USGS and Campbell.

[*Note* Campbell's later estimate of Norway's 'Regular conventional' URR is unchanged, at 34.0 Gb; while he has corresponding cumulative production to 2010 as 23.5 Gb. (Campbell 2013).]

Source IHS Energy 'PEPS' year-2000 dataset, with permission

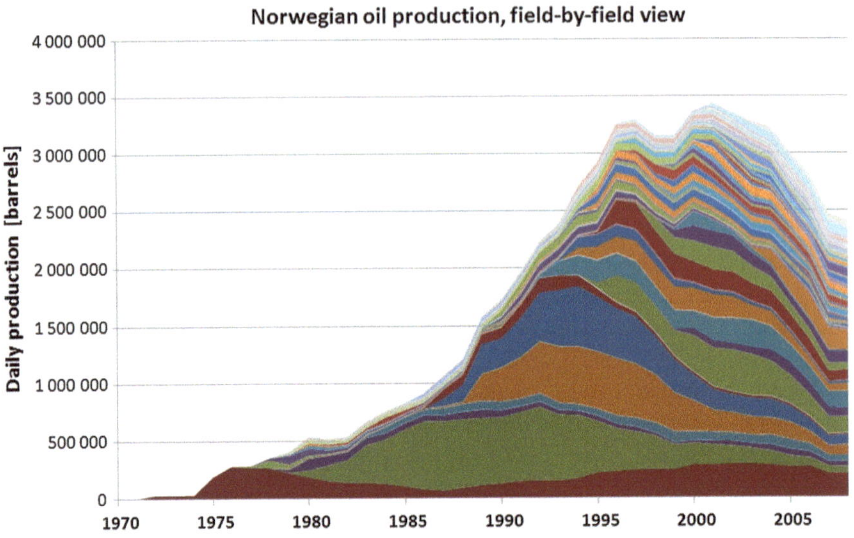

Fig. 2.13 Norway: Oil production by field, 1970–2008.
Source M. Mushalik; website: http://crudeoilpeak.info

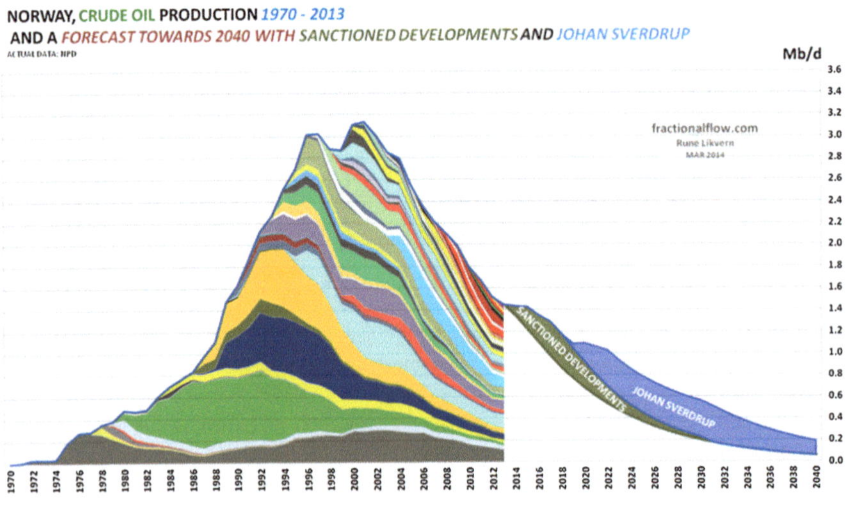

Fig. 2.14 Norway: Production, 1970–2013, and forecast to 2040.
Source Rune Likvern, (fractionalflow.com), March 2014, and see caveat on this website re possible higher future production; reproduced from M. Mushalik website: http://crudeoilpeak.info

Figure 2.13 presents Norway's production on a by-field basis, while Fig. 2.14 includes similar data plus a forecast of production to 2040.

As Fig. 2.13 shows, the individual field profiles are quite variable, and (at least for the larger fields) are less like those indicated in the simple model of Fig. 2.4 than was the case with UK field production. Nevertheless the same general mechanism driving the resource-limited peak of conventional oil production in a region is clear: decline of the large early fields not being compensated after peak by the production of the numerous later but smaller fields.

In summary, Norway is like the UK, a country where for *conventional* oil it is almost certain that only modest scope remains for new discoveries, where her original recoverable stock of this class of oil would seem to be well over half depleted, and hence her *resource-limited* production peak of this class of oil is past; Fig. 2.14.

2.3.7 Oil Discovery and Production in Indonesia, Russia and Iraq

Up to now we have looked at the production peak of conventional oil in a region via four examples: a simple model; in a small oil producer well past peak (Germany), though with enough fields to illustrate the fundamental mechanism of peak; and in two relatively large oil producers (the UK and Norway), but where their fields are totally or predominately offshore, and hence reflect fairly recent technology (that since the 1970s) from the outset.

To understand peak in a region more fully, we need also to look at large producers with a longer production history, and here we choose Indonesia, Russia and Iraq. Our research group in both the Oil Depletion Analysis Centre (ODAC) in London, and at the University of Reading, has examined the year-2000 IHS Energy 'PEPS' 2P discovery and production histories for the majority of all significant oil-producing countries, including these three, but permission to release these data has been given only for Indonesia. Nevertheless, useful observations can be made for all three countries. In each case, the main questions to ask are:

- Has the region seen one or more peaks in production?
- Were these peaks generated by politics, price, or commercial considerations; or instead by hitting the region's recoverable conventional oil *resource limit*, based what has been discovered to-date?
- Could the region see a future, higher production peak? I.e.: Do the petroleum geologists think there is potential for sufficient new conventional oil discoveries to raise future production above the earlier peak (or, equivalently, the reservoir engineers think there might be adequate gains from recovery factor increases to do the same)?

We look at these countries in turn:

Indonesia

The IHS Energy 'PEPS' data for Indonesia to 2000 are given in Fig. 2.15.

As Fig. 2.15 shows, up to the year 2000 Indonesian production had given no clear indication of peak, having remained roughly on-plateau since 1973 (and only starting to fall off post 2003). So the question is why did Indonesia not exhibit one or more peaks as in the earlier examples given above? The explanation is that the country has seen several phases of discovery: two main phases (in terms of volume discovered) in onshore oil, in the early 1940s, and in the 1970s; and a substantial phase of discovery of offshore oil, starting from the late 1960s, and tailing off from about 1995. Production thus reflected this succession of discovery phases.

So again the important question is: Is Indonesia anywhere near her resource-limited production peak for conventional oil? We can answer this from Fig. 2.15 (right). With cumulative production by 2000 at 20 Gb, and the USGS, Campbell, and extrapolation of discovery estimates for URR all agreeing

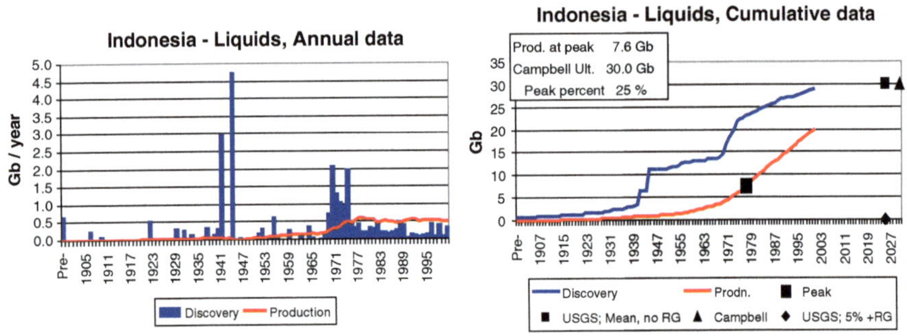

Fig. 2.15 Indonesia: Oil-industry '2P' data on oil discovery, and production; 1900–2000. Data are for the IHS Energy definition of 'liquids'; here meaning crude oil plus NGLs.

Left chart

- *Blue bars*: Annual 2P backdated oil discoveries (i.e., reflect discovery information available at 2000). Two main discovery phases are indicated: in the 1940s, and the 1970s.
- *Red line* Annual production. No significant production peak; plateau from ~1973. (As more recent data indicate, this plateau extended to 2003, following which there has been a significant decline in production.)

Right chart Cumulative plot of the same data.

- *Blue line* Cumulative backdated 2P discovery; the main two phases of discovery (the first onshore, the second a combination of onshore plus offshore) are clear from the 'steps' in the discovery curve.
- *Red line* Cumulative production. Medium-sized square indicates the approximate onset of the production plateau.

Also shown are URR estimates for Indonesia conventional oil, from the USGS year-2000 assessment (mean, no reserves growth value); and from Campbell (for 'Regular conventional' oil). These estimates, at 30 Gb, are in agreement with a rough extrapolation by eye of the discovery trend. [Campbell (2013): 'Regular conventional' URR: 32 Gb; cumulative production to 2010: 24 Gb.]

Source IHS Energy 'PEPS' year-2000 dataset, with permission

at ~30 Gb, clearly the country is *well past* her resource-limited conventional oil maximum on the basis of the 'peak at mid-point' rule.

It was this level of depletion, and hence declining production, that caused Indonesia to initially withdraw from OPEC in 2009 (though she has recently been given approval to re-join). Note that the country's resource-limit should not have come as a surprise; it could be predicted from about 1980, the date by which the combination of both onshore and offshore 2P discoveries had begun to tail off (Fig. 2.15, right), and hence the country's likely total URR could be established with reasonable confidence; and hence also the likely date of the country's *resource-limited* conventional oil production maximum.

Russia

Although no IHS Energy 'PEPS' year-2000 charts can be given here, examination of the data shows a clear drop-off for Russia in both backdated 2P discovery, and production, in the eight or so years prior to the year 2000; where these reductions were due to the collapse of the Soviet Union. Both discovery and production have recovered to some extent subsequently.

But what about Russia's peak production: Is it anywhere near? Campbell's estimate at the year 2000 of Russia's 'Regular conventional' ultimate, was 200 Gb, where this was significantly lower than either the USGS year-2000 mean estimate for Russia's conventional oil ultimate, excluding reserves growth, at ~310 Gb, and the apparent asymptote of the discovery trend at the same date, suggesting around the same figure. This difference was for two reasons: Firstly, Campbell excludes Russia's polar oil in this 'ultimate'. But secondly, and importantly, Campbell (like Laherrère and some others) judges that much of the Russia discovery data are in fact 3P (proved-plus-probable-plus-possible), rather than 2P; see Fig. 2.16.

As Fig. 2.16 shows, the TNK-BP estimate of the field's URR is much more in line with the linearised production decline trend than the ABC1 value held in some industry databases; and note that the TNK-BP estimates *includes* considerable future work to maintain field production. For additional discussion of this topic, see Laherrère (2015).

If we use the Campbell year-2000 estimate for Russia's 'Regular conventional' oil, and with her cumulative production to that date being ~120 Gb, it was clear from the 'mid-point' rule that the country could not be far from its production peak of this class of oil. This was despite the fact that many analysts at the time were predicting that Russia production would see large increases, and come to the rescue of declining production in many countries elsewhere.

Campbell's more recent estimate as of 2010 of Russia's 'Regular conventional oil' ultimate is higher, at 230 Gb; while Laherrère's estimate for Russia's all-conventional oil URR is about 250 Gb. When these are set against the country's cumulative production to 2010 of about 150 Gb (Campbell 2013, p149), Russia's peak production of both of these classes of oil would seem to be almost certainly past. And, like the previous examples given, the date of this peak could have been predicted for a long time; Russia's 2P discovery trend having started to fall from about the mid-1960s.

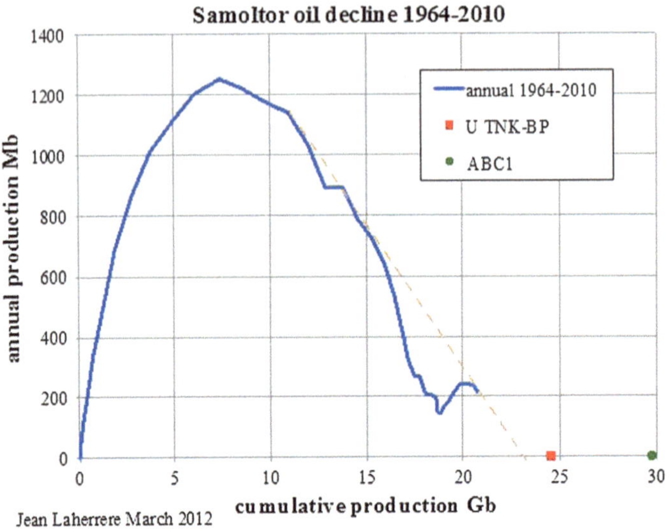

Fig. 2.16 Russia 'Linearised decline' plot of the Samotlor field: Production versus Cumulative production.

Notes

- Annual: Annual production.
- U TNK-BP: Estimate of field URR from TNK-BP.
- ABC1: Field URR held in some industry databases; often treated as proved-plus-probable (2P), but more likely proved-plus-probable-plus-possible (3P).

Source J. Laherrère, 2012

Iraq

Now we look at discovery and production in Iraq. Here the year-2000 'PEPS' data show three distinct is 'steps' of oil discovery, corresponding roughly to the major discovery plays of Kirkuk, Rumailia, and East Baghdad. The question naturally is: Does Iraq have a lot more oil to discover? Many analysts have looked to the Western desert to yield much.

But when the USGS were asked to re-visit their year-2000 assessment for Iraq in greater detail they looked thoroughly at source rocks, traps, and possible migration paths, but stuck more-or-less to their year-2000 assessment of ultimate (Ahlbrandt 2003). The mean 'non-grown' value of the latter at 145 Gb was close to Campbell's year-2000 value of 135 Gb, and also close to the asymptote of the cumulative discovery trend based on what had been found to that year. Campbell's later (2013) estimate of Iraq's 'Regular conventional' URR, as of 2010, has fallen somewhat, to 115 Gb, but this does not change the picture much.

This is because Iraq's cumulative production is still relatively small, reaching some 34 Gb in 2010 (Campbell 2013), or around 40 Gb by end-2015. So although Iraq is unlikely to have much more conventional oil than already discovered, only about one-third of this has been used so far, and hence sufficient remains to support the country's development for many years to come. However her production *peak* is probably not too far away; if her production rises fairly soon to a plateau of 6 Mb/d as some foresee, then on the simple 'mid-point' rule the country's peak of conventional oil would be in just over a decade or so.

Iraq is thus an example of a region where the conventional oil production peak, though not past, is also not so very distant. Note that given the USGS view that not much more conventional oil remains to be discovered in Iraq in new fields, her ultimate—and hence her resource-limited production peak—could have been estimated from about the last 'step' in the cumulative discovery curve, which occurred some forty years ago, around 1976.

2.3.8 *Oil Discovery and Production in Saudi Arabia*

Finally, to complete our survey of how regions behave in terms of 2P discovery and hence their production peak, we look at perhaps the most significant producer not yet past peak, Saudi Arabia. Here the relevant 'PEPS' year-2000 data are given in Fig. 2.17.

As the Fig. 2.17 (left) shows, Saudi Arabia's oil discovery is dominated by the super-giant field Ghawar in 1948, with other finds relatively considerably smaller.

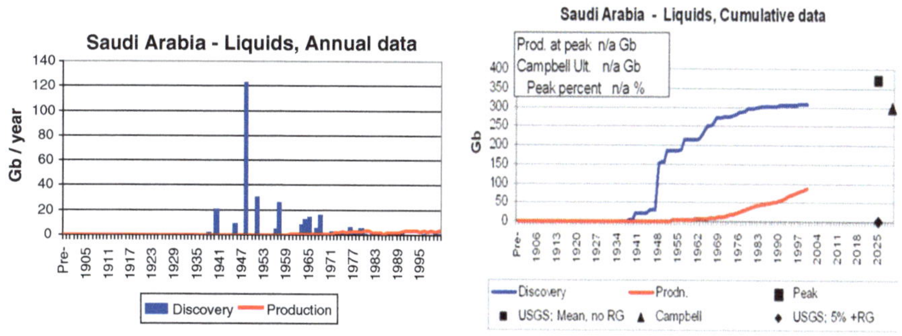

Fig. 2.17 Saudi Arabia: Oil-industry '2P' data on oil discovery, and production; 1900–2000. Data are for the IHS Energy definition of 'liquids'; here meaning crude oil plus NGLs.

Left chart:

- *Blue bars* Annual 2P discoveries (data are year-2000 backdated, so reflect information available at this date). The earliest find was in 1938 at the Dammam well No. 7 (now modern day Dahahran); but the country's super-major discovery was Ghawar in 1948, with relatively little discovered since 1970.
- *Red line* Annual production. Early peak at 10.3 Mb/d in 1980, and rough plateau from 1990. [More recent data show the absolute peak in production to-date is in 2015.]

Right chart:

- Cumulative plot of the same data, plus two estimates made around the year 2000 of Saudi Arabia's conventional oil 'ultimate'. [*Note* Campbell (2013): 'Regular conventional' URR estimate unchanged at 300 Gb; cumulative production of this class of oil to 2010: 117 Gb.]

Source IHS Energy 'PEPS' year-2000 dataset, with permission

Production has been variable over time, mainly due to the imposition of quotas. The questions we are seeking to answer, of course, are: What is the country's likely future production; and how close is peak?

To answer these we first need to know how well the simple model of Fig. 2.4 applies to the large Middle East producers such as Saudi Arabia. A number of people, including for example Dr. Adnan Shihab-Eldin (former Acting Secretary General for OPEC), have questioned the applicability of the simple model of Fig. 2.4, which captures the main drivers for the peak in countries like the UK, to large Middle East producers. This is a sensible question, as the latter have oil discovery and production characteristics quite a bit different to countries like the UK. Specifically, in the large Middle East producers:

- There is usually one, or a small number of, extremely large fields, and then a succession of smaller, more typically-distributed fields in terms of size.
- This one, or a few, extremely large fields have normally been held on-plateau for long periods (rather than rising fairly quickly to peak production and the declining); but also, from time to time, they have seen wide excursions in production resulting from OPEC quotas, or civil or military unrest.
- Some of these countries, and Saudi Arabia especially, have considerable amounts of oil remaining in untapped ('fallow') fields, waiting to come on-stream once the very large fields go into decline.
- Finally, oil exploration in these countries is financially different from those where commercial companies pay for exploration. This is because, since expropriation, exploration expenditure and field upgrades in these large producers often has to be paid for in real funds, not as '10-cents-on-the-dollar' tax-deductible expenditure as was the case when the oil companies were in control.

To see if the simple model of Fig. 2.4 applies to a producer such as Saudi Arabia, we need to look in detail at the country's typical field profiles; and also at the estimates for URR. We start with field profiles.

Field profiles:

We can examine the country's field profiles via two publically-available bottom-up by-field forecasts for Saudi Arabian production dating from 2008, given in Figs. 2.18 and 2.19.

Of particular importance in these forecasts is the expected production profile of Ghawar. This field has seen reportedly excellent work in terms of water injection wells along the field flanks, and one reservoir engineer familiar with the field suggests that production will 'go out like a light' at the end of this injection phase. However, the field is far from homogenous along its length, and both the above two forecasts indicate instead that Ghawar's production will tail off over time in a typical exponential fashion.

In the Energyfiles Ltd. forecast, other fields are also shown with exponential decline in output; this is in contrast to the Miller forecasts (over a shorter period) which shows the *large* other fields as holding a relatively flat output to the end of the forecast period, with smaller fields (and the yet-to-find fields) showing typical field decline.

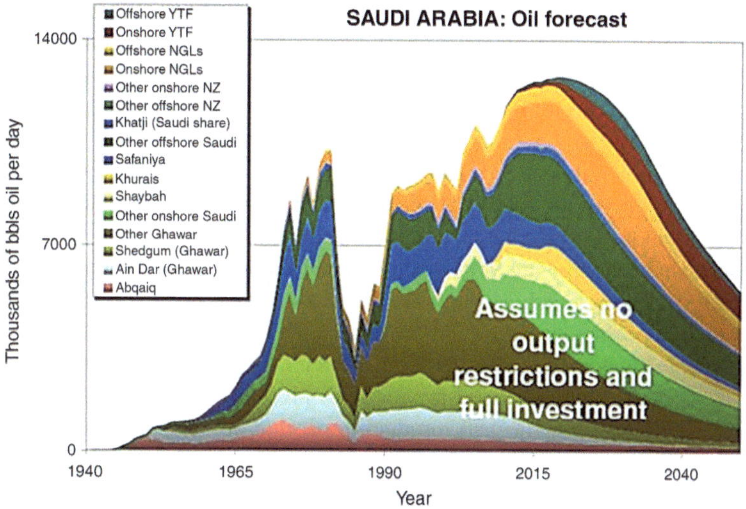

Fig. 2.18 Saudi Arabia: Forecast, made in 2008 by Energyfiles Ltd., of production by field, and field type, to 2050.

Note Ghawar production is shown as three separate components.

Source Annex of *Technical Report 7,* UKERC *Global Oil Depletion* study, 2009

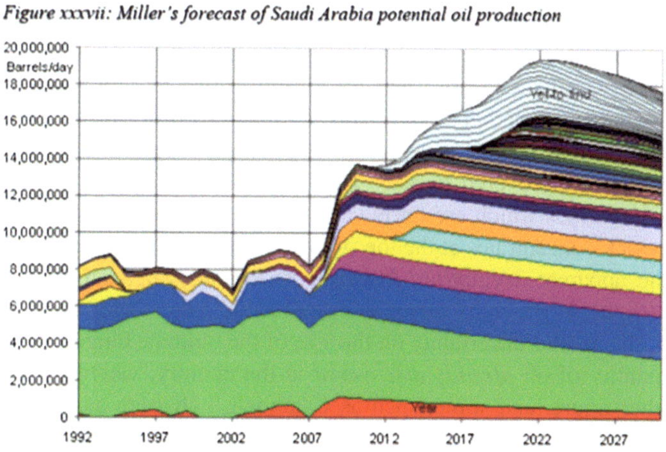

Fig. 2.19 Saudi Arabia: Forecast, made in 2008 by R. Miller, of production by field to 2030.

Note Ghawar production is represented by the *light green* field *second from bottom.*

Source Annex of *Technical Report 7,* UKERC *Global Oil Depletion* study, 2009

In terms of total output, the two forecasts give significantly different results: Energyfiles predicts a peak in production at about 12 Mb/d occurring in about 2020; while Miller forecasts a much higher peak, at nearly 20 Mb/d, occurring around the same date. The explanation for the difference is that Miller's forecast includes the 'Miller bump' in production, due to bringing on rapidly all the country's currently fallow fields (UKERC 2009). But Miller reports (private communication) that he thinks such a situation unlikely, and would probably support the more cautious view shown in the Energyfiles' forecast.

The main conclusion from these two forecasts is that the underlying mechanism of Fig. 2.4 of resource-limited oil peaking in a region—the result of adding the output from large early fields, and then from smaller fields—is expected to operate in the same fundamental way for the large Middle East producers as for countries already past peak, such as the UK. There is a caveat however: with their long flat production profiles of large fields, regions such as these very large producers would be expected to see peak later, in terms of percentage of URR, than is typical for the more normal 'unconstrained-production' producers, such as the US, UK, or Norway; so resulting in peak production *at mid-point of the URR or a bit after*, rather than more typically for other regions of peak at mid-point or a bit before. (Incidentally, for forecasts for Saudi Arabian production by the IEA and EIA made at the same date see Annex 5.)

Next we look at the second part of the puzzle for predicting Saudi Arabia's output: estimates of her likely conventional oil URR.

What URR to assume for Saudi Arabia?

There is disagreement (indeed, true uncertainty outside of the country) on what is Saudi Arabia's likely URR for conventional oil.

Though some authors foresee very large volumes of oil yet to be discovered in the country, based in part on the relatively few exploration wells (only somewhat under 200 'new-field wildcats') that have been drilled to-date; among petroleum geologists there is—perhaps surprisingly—only modest uncertainty over the country's yet-to-find quantity, where this is generally taken as not especially large compared to that discovered to-date. This is due to the region's geology of large salt-sealed anticlines which are fairly easy to map via seismic, and where oil from the source rock concentrates at the top of these, with little oil expected between. As a result the apparent asymptote of discovery to-date, for example, as shown in Fig. 2.17 (right), is generally taken as a reasonable indicator of URR.

Instead, the greater uncertainty on the size of the ultimate hinges on knowing the true 2P quantity of oil *already discovered* in the country; where some authorities take a higher figure, and some, such as Campbell, Laherrère and others, a lower. The explanation for this lies in Fig. 2.20.

Figure 2.20 shows several trajectories for the apparent Saudi Arabia 2P cumulative discovery curve from an industry source, as generated at five different dates spanning 1998–2011. Specifically for example, the curve at 1998 indicates an 'extrapolated-by-eye' URR of ~300 Gb, while the three curves from 2005 and later indicate the equivalent URR at about 400 Gb or a bit higher.

The curves thus provide (a rather extreme) example of one aspect of apparent 'reserves growth' in some of the industry data. While many cases of reserves growth in a field or region can be a significant, due to improved technology or an

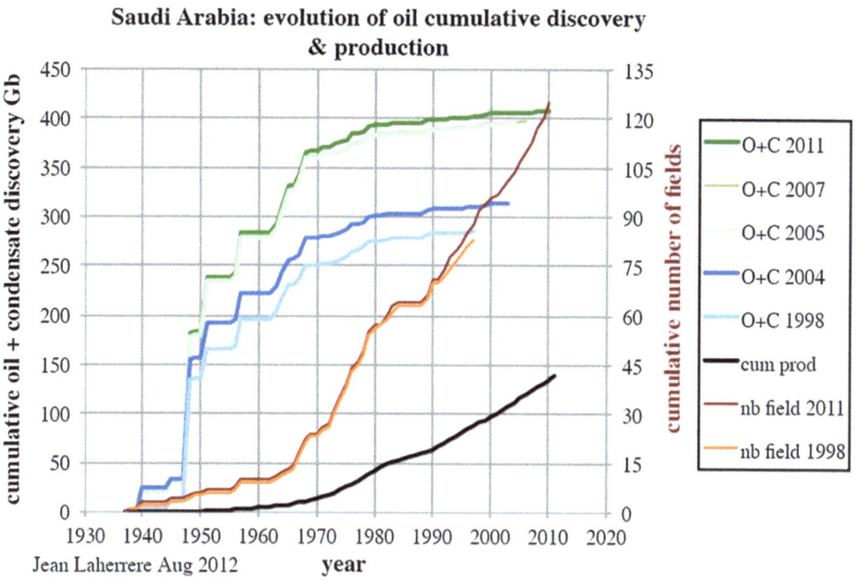

Fig. 2.20 Saudi Arabia: Backdated notionally '2P' cumulative oil discovery data from an industry source, as reported over a number of different years spanning 1998–2011; also cumulative production (*both left-hand scale*); and cumulative number of fields (*right-hand scale*).

Legend
- O+C: Oil plus condensate, plus year when data reported.
- cum prod: Cumulative production to 2011.
- nb field: Number of fields discovered, plus year when data reported; right-hand scale.

Source J. Laherrère, Aug. 2012

increase in oil price (see discussion in Annex 4, section 3), and hence need to be taken into account in oil forecasting, in this case Laherrère has a different explanation for the large gain in apparent URR. He notes that IHS Energy is possibly now obliged to report Saudi Aramco official reserves estimates for the country, such that IHS Energy's cumulative discoveries for Saudi Arabia currently stand at about 400 Gb as shown in Fig. 2.20; and hence in agreement with the country's reported end-2011 remaining *proved* (1P) reserves of 265 Gb, plus cumulative production to the same date of ~140 Gb. Previously Petroconsultants (IHS Energy's predecessor company) had reported a much lower number for Saudi Arabia cumulative 2P discovery, closer to the 300 Gb shown in Fig. 2.20 for the 1998 data. Also shown on this plot are estimates for the number of fields at two different dates, indicating that it was not the addition of new fields that explains the discrepancy. Further details supporting this view from Laherrère are given in Laherrère (2015).

So what URR should we use for estimating Saudi Arabia's date of conventional oil peak? Figure 2.17 (right), vintage 2000, indicates the USGS year-2000 estimate for URR (mean, no reserves growth, ex-NGLs) was ~370 Gb; while Campbell's estimate of the same date for 'Regular conventional' oil (ex-NGLs) was 300 Gb.

Since the USGS year-2000 estimate was made on the basis of 1994 Petroconsultants' data, and hence was quite a bit above the 2P discovery trend at that date, the argument was possibly made that Saudi Arabia had had no great need to explore for more oil while quotas were in place. Campbell's subsequent 2010 data (Campbell 2013) still holds the URR at 300 Gb (ex-NGLs), while the cumulative production (also ex-NGLs) to that date was 117 Gb. (This contrasts with the country's cumulative production to 2011, including NGLs, being ~140 Gb as shown in Fig. 2.20) Adding another 5 years of production to the 2010 data puts Campbell's cumulative production today at ~135 Gb, indicating, if the mid-point rule is used, that Saudi Arabia's peak would be in only four years or so; although as explained above, for large Middle East producers it may be reasonable to expect peak somewhat later than mid-point.

Nevertheless, this provides a warning of the coming supply difficulties. If instead, the 'including NGLs' cumulative production to-date value of 155 Gb is used, and this is combined with an assumed higher URR (incl. NGLs) of, say, 430 Gb, then Saudi Arabia's 'mid-point' peak is expected in 2030; but see also the by-field forecasts of Figs. 2.18 and 2.19 which put the production peak earlier, around 2020 or so.

In either case, note that Saudi Arabia's backdated 2P cumulative discovery data had substantially flattened out by about 1970 (Fig. 2.17 right), allowing reasonable forecasts of her likely date of production peak to be made from that date, provided a view was taken on the accuracy of the 2P discovery data used.

2.3.9 Summary of Findings on Peak Conventional Oil Production in Regions

In summary, in this section on explaining the peak of conventional oil production in *a region*, we have examined the mechanism for peak as indicated by a simple model, and looked at past or future peaks in a number of counties as follows: Germany (a small producer well past peak); the UK, Norway and Indonesia (medium producers past peak); Russia (a large producer at or close to peak); and Iraq and Saudi Arabia (large Middle East producers whose production peaks are in the future, though not so very far away). Table 2.1 summarises these findings, and includes data for the US and the world, as supplied elsewhere in this book.

Despite these differences in their dates of peak production as given in this Table, the common factor of all these countries—and indeed of virtually all regions globally—is that their dates of peak *discovery* of conventional oil (oil in fields) are well past, as indicated by the inflection points in their backdated cumulative 2P discovery curves.

This in turn permits the simple approach to be applied of extrapolating this discovery trend (combined with geological knowledge to indicate possible new

Table 2.1 Dates of Peak 2P discovery of conventional oil compared to peak in production

	C'bell ~ 2000 (Reg. cv. oil)	USGS 2000 (Cv. oil ex-NGLs)	C'bell 2010 (Reg. cv. oil)	C'bell cum. prodn. to 2010 (Reg. cv.)	Cum. prodn. to end 2015 (Reg. cv.)	C. Pr. 2015 as % C'bell URR	Main oil loc'n.	Date of peak discvy. (All-cv. oil)	Date of peak prodn. (All-cv. oil)
	URR	URR	URR						
	(Gb)	(Gb)	(Gb)	(Gb)	(Gb)	(%)	(On/off)	(Year)	(Year)
Notes	*1*	*2*	*3*	*4*	*5*	*6*	*7*	*8*	*9*
Country									
Germany	2.4	2.1	2.5	1.96	2.05	82	On	1950s	1967
UK	32	29	32	24.7	26.3	82	Off	Mid-1970s	1999
Norway	34	35	34	23.5	26.9	79	Off	1970s	2001
Indonesia	30	30	32	24.0	25.6	80	All On Off	'40s, 70s '40s, 70s '60s–95	1977[a]
Russia	200	303	230	150	165	72	On	1960s	1987[b]
Iraq	135	145	115	34	40	35	On	Var.	–
Saudi Arabia	300	371	300	117	135	45	On	1948	–

(continued)

Table 2.1 (continued)

	C'bell ~ 2000 (Reg. cv. oil)	USGS 2000 (Cv. oil ex-NGLs)	C'bell 2010 (Reg. cv. oil)	C'bell cum. prodn. to 2010 (Reg. cv.)	Cum. prodn. to end 2015 (Reg. cv.)	C. Pr. 2015 as % C'bell URR	Main oil loc'n.	Date of peak discvy. (All-cv. oil)	Date of peak prodn. (All-cv. oil)
	URR	URR	URR						
	(Gb)	(Gb)	(Gb)	(Gb)	(Gb)	(%)	(On/off)	(Year)	(Year)
USA	260	286	200	179	194	97	On	Mid-1930s	1970
World	~2000	2307	2000	1093	~1220	~60	On	1960s	2005[c]

Also shows: Cumulative production of 'Regular conventional' oil to end-2015 versus Campbell's 2010 estimate of URR for this class of oil

Notes

1: Campbell 'Regular conventional' oil is 'All-conventional' oil less: deepwater, polar, very heavies, NGLs. *Source* Personal communications and Campbell publications

2: USGS Year-2000 Assessment data: Mean values, ex-NGLs, no adjustment for reserves growth. Source: USGS (2000)

3: *Source* Campbell (2013)

4: *Source* Campbell (2013)

5: Production data from BP *Stats. Rev.*, for 2011–2014, plus for 2015 estimated; and multiplied by a guessed 80 % if the country or region in question is currently producing significant quantities of non-'Regular conventional' oil

6: Cumulative production to end-2015 of 'Regular conventional' oil (estimated as above) as a percentage of the Campbell 2010 URR estimate for this class of oil

7: Main location of the country's oil, onshore or offshore

8: Date of peak '2P' discovery of 'All-conventional' oil

9: Date of peak production of 'All-conventional' oil

[a] Indonesian production was roughly on plateau: 1977–1996

[b] Date of this Russian peak was prior to the economic collapse of the FSU, but Russia's recoverable conventional oil resource base is now such that a significantly higher future peak in production is no longer possible

[c] This is the peak date of global 'Regular conventional' oil. The global peak date of 'All-conventional' oil is less certain, see text

For more recent USGS data, see Annex 4

plays) to estimate each region's likely conventional oil URR; and hence using the approximate 'mid-point' rule to predict the date of peak production.

Now we turn to a different and important question: how to predict the date of peak conventional oil production *for the world as a whole.*

2.4 Peak of *Global* Conventional Oil Production

Predicting peak conventional oil production for the world is different to predicting peak in a region. If a given region peaks, the world's supply of oil can be met by production ramping up in other regions, as has happened many times in the past; most notably when US Lower-48 conventional oil production peaked in 1970, with enough production to more than compensate coming from new, but already-discovered, regions such as Alaska, the North Sea, and in Mexico, Russia and elsewhere. To understand the global peak of conventional oil two new factors need to be considered: the comparative *cost* of conventional oil production from different regions and for different classes of oil, and also the global *price* of oil.

The comparative costs are needed because as one region or class of oil goes into decline (or is restricted into the market for political or other reasons), conventional oil from other regions or classes may replace it; the price of oil needs to be known to see if this is high enough to support such marginal production.

2.4.1 IHS Energy Year-2000 Global Discovery and Production Data

First however, in looking at the data for the global peak we start with the familiar IHS Energy year-2000 backdated 2P discovery and production plots; here for the world as a whole, as given in Fig. 2.21.

To understand the background to the global peak, these plots are worth examining in detail.

In the left plot the discovery of Ghawar in 1948 is clearly visible, as is the bulk of global conventional oil discovery which occurred in the period from 1950 to 1990, where this resulted from increasing knowledge of the geology and physics of 'oil systems', and from the widespread use of seismic techniques (and specifically digital seismic from about the mid-1960s).

But the plots also clearly set out the global availability of oil throughout this century. As the left plot shows, certainly from about 1925 onwards and probably before, the rate of discovery of conventional oil ran well ahead of oil production, and as a consequence this put a large quantity of oil in the 'global oil bank' in the form of 2P reserves. (Recall that this left plot is 'to-scale', in the sense that the total

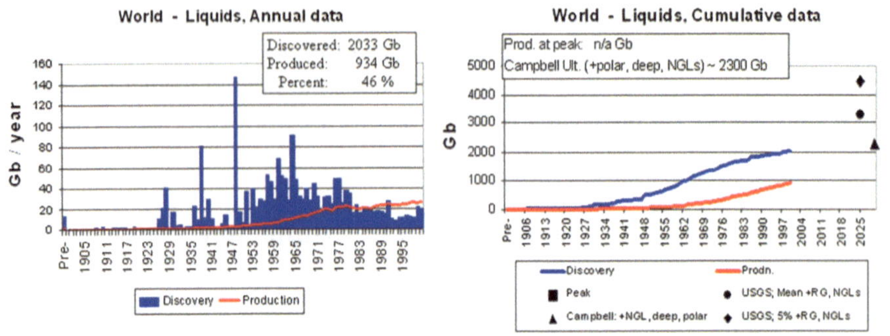

Fig. 2.21 World: Oil-industry '2P' data on oil discovery, and production; 1900–2000. Data are for the IHS Energy definition of 'liquids'; here meaning crude oil, NGLs, tar sands oil and Venezuela heavy oil. (The IHS Energy definition includes shale ('light-tight') oil and also oil from kerogen, but relatively little of these classes of oil were produced up to 2000.)

Left Annual backdated 2P discovery: *vertical bars* Annual production: line.

Right Same data on a cumulative basis.

Data from IHS Energy year-2000 'PEPS' database, with permission

area covered by the blue bars by a given date measures oil discovered in the same units as the area under the red line gives total oil consumed by the same date.)

This happy supply situation started to change at the date at which the rate of conventional oil being discovered peaked, and then started to decline, as indicated by the inflection point in the discovery curve in the right (cumulative) plot. As this shows, the *rate of global discovery of conventional oil in new fields peaked in the mid-1960s*; a key fact, still not sufficiently known, in understanding future global oil production. But, even so, in the mid-1960s, annual discovery, though it had peaked, was still running ahead of annual production, so the 2P reserves 'in the bank' were still increasing, although progressively more slowly.

This global oil supply situation changed yet again in about 1980, the critical date at which global production finally caught up with rate of global discovery of oil in new fields (see the left plot). At this date the 2P global reserves 'in the bank' started to decline. But as the right plot shows, even by the year 2000 there was still a large quantity of these reserves remaining; at just over 1000 Gb, representing about 40 years' worth of global production at that date.

With this information in hand, we can now address the question of the approximate date of the global production peak of conventional oil.

The right plot of Fig. 2.21 shows three estimates from the year 2000 of the world's conventional oil URR (but note that the definitions of these estimates are different to those in the similar previous by-country plots). These estimates are:

- Campbell's year-2000 estimate for global 'Regular conventional' oil, but here also *plus* deepwater and Arctic oil, plus NGLs; URR ~2300 Gb (triangle, at ~2030).

– USGS year-2000 mean estimate for the global URR of 'all conventional oil' (including NGLs) and here *including* upward adjustment for reserves growth; URR = 3345 Gb (circle, at ~2025).
– USGS year-2000 corresponding '5 % probable' estimate (again including NGLs plus adjustment for reserves growth); URR ~4400 Gb (diamond, at 2025).

Notes:

– Campbell's URR would seem to be below the 'extrapolated 2P discovery' line. This is because as mentioned earlier he, like Laherrère, discounts some of the OPEC and FSU oil in the industry datasets.
– Both the USGS year-2000 URR estimates shown include reserves growth, the USGS in its Year-2000 Assessment having applied reserves growth functions globally for the first time (see Annex 4). While such growth is certainly technically possible, as can be seen from this plot these 'assessed' URR estimates were out of line with extrapolation of the global 150-year 2P backdated discovery trend. For such very large extra amounts of conventional oil (~1000 Gb in the 'mean' case; and ~2000 Gb in the '5 %' case) to become available early enough to avert a global peak appeared very unlikely to many observers at that date.

On these data, it was already clear in 2000 that the World was close to its peak of conventional oil production if the 'peak at mid-point of URR derived from extrapolated 2P discovery' rule was applied. Extrapolating the cumulative 2P discovery and production lines forwards indicated that the global peak of conventional oil production was likely to occur around the year 2010. Indeed, as the Figure shows, based on the peak of discovery in the mid-1960s it had been clear for a long time that the global peak of conventional oil was likely to occur roughly around the year 2000, or not long after.

2.4.2 More Recent Data on Global 2P Oil Discovery and Production

Now we look at more recent data for global backdated 2P oil discovery and production. Figure 2.22 gives these data as supplied by Jean Laherrère for crude oil, but excluding NGLs and extra-heavy oil (the latter mainly tar sands and Orinoco oil); and also the data for gas.

As the Figure shows, Laherrère's 'exploration geologist' view of the likely extrapolation of the backdated cumulative 2P discovery curve indicates a 'medium-term' URR for global 'conventional' oil (crude less extra heavies, less NGLs) of ~2200 Gb. On this basis, and using the 'peak at ~ mid-point' rule, and comparing to cumulative production shown for this class of oil to end-2013 of ~1230 Gb, the 'expected' date of peak for this oil would have been ~2005; in general agreement with the apparent actual date, see Chap. 4.

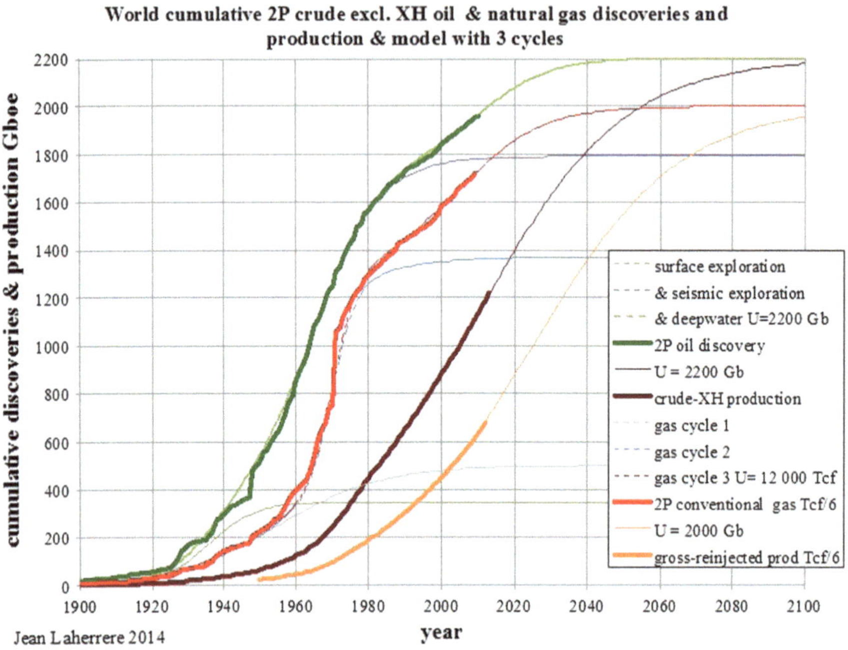

Fig. 2.22 World: Cumulative 2P Backdated Oil Discovery 1900–2010, and forecast to 2100; Cumulative Oil Production, 1900–2013, and forecast to 2100. (Also shown are the corresponding discovery and production data for gas.)

- *Leftmost line* Laherrère's judgement of 'most probable' backdated 2P cumulative global discovery data for crude oil plus condensate, less extra heavy oil (the latter mainly Athabasca tar sands and Orinoco oil), and not including NGLs.
- *Next left line* Corresponding data for gas, calculated as Tcf/6.
- *Next leftmost line* Cumulative global production of crude oil less extra heavy oil and NGLs.
- *Rightmost line* Cumulative global production of gas, Tcf/6.

Laherrère writes: 'The 2P discovery data reflect data from industry 'scout' sources, but reduced by: 300 Gb to allow for overstatement of the OPEC Middle East original reserves data (as confirmed by Sadad Al-Husseini, former VP Aramco, 2007 Oil & Money conference London); by 30 % of the FSU data (~ 100 Gb) to allow for the datasets ABC1 holding probably closer to 3P than 2P data (as indicated by field decline plots, see e.g., Fig. 2.16, and by Gazprom audits in annual reports); and by 200 Gb to allow for Orinoco 2P discovery data reflecting non-conventional oil.'

Source J. Laherrère

Alternatively, one can use the PFC Energy 'peak at 60 % of discovered' rule, and estimate the global peak date for conventional oil ex NGLs. On the basis of the data shown here, 60 % of current 2P discovery (at 1950 Gb) is ~ 1170 Gb, which predicts peak somewhat later, at ~ 2011–2012.

In comparison to Fig. 2.22, Fig. 2.23 gives the corresponding IHS Energy data for global oil discovery and production up to 2011 for the company's 'Liquids' category, which includes NGLs, light-tight oil, extra-heavy oil (the latter mainly tar sands and Orinoco oil), and oil from kerogen, but excludes GTLs, CTLs and biofuels.

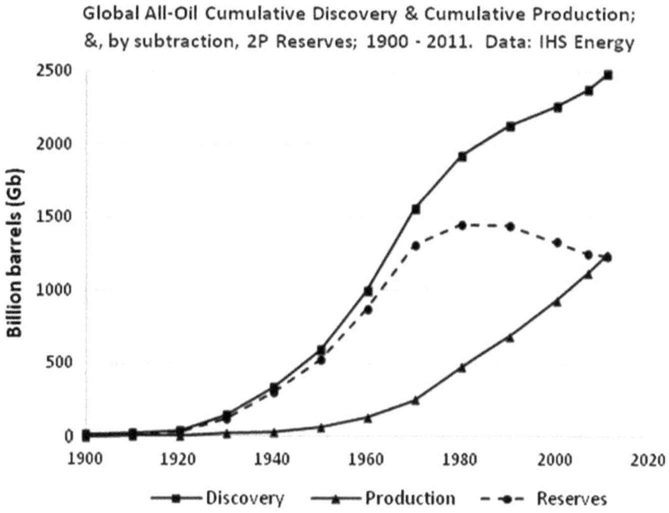

Fig. 2.23 World: Cumulative 2P backdated oil discovery, and cumulative oil production (also hence 2P Reserves by subtraction), 1900–2011.

Notes

- The plot shows IHS Energy 'Liquids' data, stated to include: "crude oil, condensate, NGLs, liquefied petroleum gas, heavy oil and syncrude". The data thus include light-tight oil, and oil from tar sands and Orinoco oil, but exclude GTLs, CTLs, biomass, and refinery gain.
- The plot is generated by reading data at 10-year intervals from Fig. 7 of Miller and Sorrell (2014) for cumulative discovery from 1900 to 2007, and from the corresponding Fig. 3 for cumulative production over the same period. Included in this plot are the data for end-2011 as given in the text of the Miller and Sorrell paper.
- Data are 2P, except for the US and Canada non-frontier areas, where the data are proved ('1P') data. The 2P data are backdated, in that they reflect information available to the IHS Energy as of 2007 (for the discovery curve), and to 2011 (for the final discovery data point). Reserves are calculated here (as done also by IHS Energy) by subtracting cumulative production from cumulative discovery.
- IHS Energy data are for oil in *fields* for conventional oil; and as announced in *projects* for non-conventional oils. The 'up-tick' in global discovery of this 'all-oil' visible from about the year 2000 (and hence the slowing in the fall-off of 2P reserves) is due to increasing inclusion of data for tar sands projects, and subsequently for US shale (light-tight) oil projects. Data are hence largely for conventional oil up until about the year 2000, after which significant amounts of tar sands and Orinoco projects were included, and most recently also data for 'light-tight' oil projects.
- As the plot shows, the global proved-plus-probable (2P) all-oil reserves at end-2011 were ~1250 Gb. This contrasts to the corresponding end-2011 value for global *proved* only (1P) all-oil reserves (from *BP Stats.*) of 1652 Gb. The difference is partly the amount of non-conventional oil included in the two sets of reserves figures, but is mainly due to the likely overstatement of Middle East OPEC *proved* reserves

In Fig. 2.23 the global discovery of *conventional* oil (incl. NGLs) might be judged (based on the pre year-2000 trend) to be heading for an asymptote URR around 2500 Gb, and where hence the production mid-point, and hence production peak, of this value is around 2011. As noted, the reasons for the difference in the data between Figs. 2.22 and 2.23 are due to the inclusion of different categories of oil, and to Laherrère's view on the need to pull down industry 2P discovery data for the FSU, some Middle East countries, and Venezuelan Orinoco oil.

Note that on the Fig. 2.23 data, as in Figs. 2.21 and 2.22, one can see that the rate of global discovery of conventional oil peaked in about the mid-1960s; and the volume of global 2P reserves in about 1980, the latter at about 1450 Gb if NGLs are included.[26]

Now we turn to the first of the two additional issues that bear on the date of *global* peak that were identified earlier: the comparative costs between different regions and classes of conventional oil. The data for this are given in Fig. 2.24.

2.4.3 Recoverable Volumes Available of Different Types of Oil as a Function of Cost

Figure 2.24 is from the IEA and gives estimates, as of 2013, of the remaining quantities of oil available from different categories of oil versus their production cost ranges.

As Fig. 2.24 shows, and as would be expected, conventional and non-conventional oil from different sources come at different ranges of costs. The first six categories shown (up to 'Arctic'), and also 'Ultra-deepwater', refer to mainly conventional oil as we define here. The other five categories are either classed here as 'other-oils' (Extra heavy, Light tight and Kerogen), or as 'other liquids' (GTLs & CTLs). The plot includes only fossil resources, so does not include biofuels.

In an ideal free market all the cheapest oil shown in Fig. 2.24 would be produced first. In practice in some regions, especially in some MENA and other countries following 'resource-nationalism' policies, conventional oil production has been limited by restricting international exploration or production access, or held back to support a higher oil price, or kept in reserve "for our grandchildren". Also at times in these and other countries oil has not been available due to strife or war. As a result of these 'above-ground' factors, conventional oil production peaks have occurred in some of the higher cost oil regions earlier than would have been the case.

Nevertheless, despite such effects, charts like Fig. 2.24 have long seemed reassuring to many oil analysts. For example, the UK's Department of Trade & Industry refuted talk of a global peak by pointing out—based on a 2008 version of this chart—that: "There is more than enough oil available to meet foreseeable demand". This view might seem natural enough; after all this chart shows ~7000 Gb of recoverable oil of all types remaining, with a century-and-a-half of production having produced only ~1250 Gb. But this view is naïve. The correct way to read a chart like this as follows:

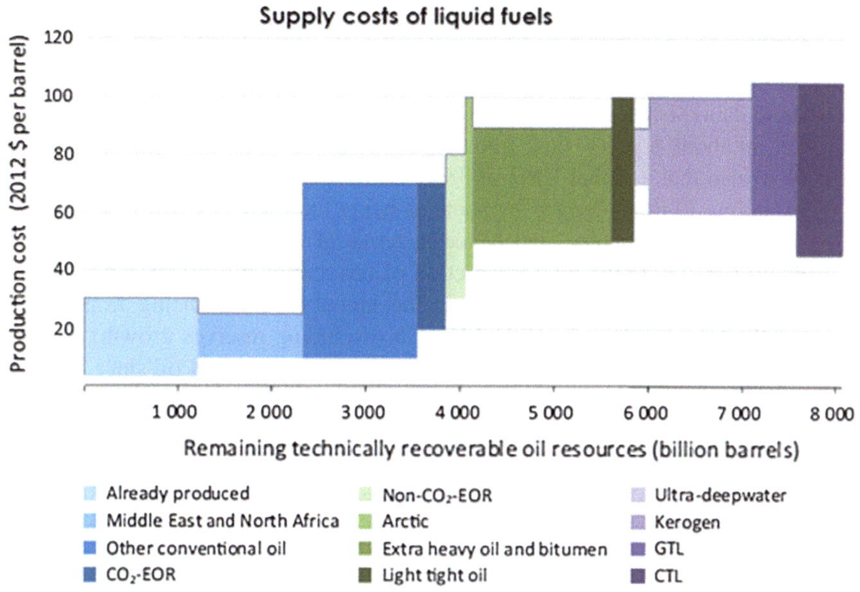

Fig. 2.24 Estimated global remaining technically recoverable volumes of oil available, by category (in Gb), versus Production cost range (in \$2012/bbl).

Notes

- EOR: Enhanced oil recovery; *MENA* Middle East and North Africa; GTL: Gas to liquids; CTL Coal to liquids.
- Volumes of oil potentially available are shown by length along the x-axis, not by the area indicated.
- [Incidentally that there are two types of chart that indicate production cost of oil versus category. One is this type, where volumes of recoverable oil are indicated together with their production cost ranges. The second type shows global oil production at given date by category versus cost make-up. An example by IHS-CERA assuming a 15 % rate of return for investment is given in Miller and Sorrell (2014). This indicates that current oil production costs range from \$22/bbl for Middle East oil to \$160/bbl for 'upgraded Canadian tar sands mine' oil. Some estimates put current Middle East production costs as significantly higher than these IHS data. In addition, other studies look at the *price* needed from oil exports to balance national budgets of major exporters; these suggest that the 'financially required oil prices' are currently around \$100/bbl and above.[4]]

Source IEA *Resources into Reserves*, 2013

(a) Understand 'mid-point' peaking of conventional oil production, so on the data shown here expect the global peak of conventional oil when ~1925 Gb has been produced (i.e., half of the ~3850 Gb URR resulting from summing the 'Already produced,' MENA, Other conventional, Arctic, plus Ultra-deepwater; and excluding EOR as this usually comes on only late in a region's life). At the current production rate of ~30 Gb/yr., and with 1250 Gb already produced by 2012, the global all-conventional oil 'mid-point' peak is expected roughly 20 years from this date, depending on the rate of demand growth.

(b) Then recognize that production of much of the MENA oil will probably not increase significantly—despite current trends—for resource-national reasons.
(c) So look instead for the 'mid-point' of total non-MENA conventional oil. On these data this occurs when 1370 Gb has been produced (half of ~2740 Gb); i.e., in about 4 years' time from 2012.
(d) Note also that a global URR of 3850 Gb for conventional oil (incl. NGLs but excluding EOR) is judged by some as being too high, at least if compared to extrapolated discovery, and hence in terms of the likely date of peak. Such analysts estimate the total production of conventional oil (incl. NGLs) out to the year 2100 (i.e., approximately the global URR) as being from 500 to 1400 Gb lower than the IEA' 3850 Gb (excluding reserves growth) number.
(e) Hence conclude, correctly, that the steep rise in the price of oil since 2004 has been driven by the slowing increase in global *conventional* oil production resulting from proximity to its peak, which in turn has forced the world to obtain the marginal barrels needed to meet growing oil demand from the expensive oils shown to the right of the graph.
(f) And, finally, recognise, that these more expensive oils also tend to have poor energy returns on energy invested (indeed a major factor in why they are costly to produce); and can face other constraints to their production such as permitting, water requirement, CO_2 emissions, and volume of waste if produced by mining.

Next we look at the second factor mentioned above that needs consideration when predicting the global peak date of conventional oil production, that of the price of oil itself.

2.4.4 Impact of the Price of Oil on the Availability of Oil

The price of oil is important because as this rises exploration is encouraged and oil that was previously uneconomic can be brought to market. Economists have made much of this dynamic, though usually citing the flawed apparent replacement of the proved reserves data to prove their case (see Chap. 5). The proper question is: At a given oil price *by how much* can conventional oil production increase?

The detailed answer must come from reservoir engineers; and is also a function of date as technology advances with time. Some answers *by field* have been illustrated above, and we have also partial answers at least by region (and also globally) from the roughly decade of high oil prices (above $50/bbl in today's money) from 1974 to 1985, and a second decade of prices above this level from 2005 to 2015. Based on these data—for example, for US Lower-48 production over the first period, and the UK production for the current period—the answer is '*by not much*', though admittedly the economists may have a valid point in saying that a decade is probably not enough for new exploration plays and production paradigms to come fully into effect.

Overall however, on the above field data and regional data, it would seem unwise to expect a high price, even $100/bbl, to bring on very much in the way of extra conventional oil. Moreover, the price cannot go too high: as we know fairly solidly from both the earlier and the current period of high oil prices, that a price much above $50/bbl in today's money slows economic activity, and reduces oil demand at least in the developed countries; while above about $100/bbl in real terms it would seem to lead to global recession.

2.4.5 Summary of Global Peak of Conventional Oil

In summary, in this section examining the likely date for the peak of global conventional oil production, we can say:

- This date is not yet known for certain.
- In part it depends on:
  - how conventional oil is defined; e.g., with or without: NGLs, EOR, heavy oil (other than tar sands) needing thermal treatment, etc.;
  - the extent that the small number of mostly Middle East 'swing' producers decide (or not) to rein in their potential production to 'save oil for the grandchildren';
  - the extent and rate that currently fallow fields are brought on-stream;
  - and the impact that a sustained high oil price (of, say, >$100/bbl) would have on the use of in-fill drilling, extra use of EOR, etc.
- But as Figs. 2.21–2.24 show, and also the forecasts of Chap. 4, the *date for the global peak of conventional oil production cannot be too far into the future.*

2.5 Peak of the Global Production of 'All-Oil', and 'All-Liquids'

So now we must turn from examining the date of peak of global *conventional oil* production to the peak for the global production of *all-oil*; and also that of *all-liquids*, where the latter includes GTLs, CTLs and biofuels.

Here a key point from the IEA data in Fig. 2.24 is that as the global peak of conventional oil production approaches, and hence the scope for increases from its production starts to tail off, the world is forced to use increasing volumes of the non conventional sources to satisfy any increase in demand. And since, as indicated in the Figure, these sources are generally more expensive to produce than conventional oil, the overall oil price is expected to rise to the level of the marginal barrel required. As mentioned earlier, this marginal-barrel price is currently roughly around $100/bbl, although the highest IHS-CERA value given in the Miller and

Sorrell (2014) paper quoted earlier is ~$160/bbl for 'Canada oil sand mine upgraded' oil assuming a 15 % rate of return.

At some later date, once the peak of global conventional oil production is past and the decline in the production of this oil has settled in, modelling shows that global production of this oil is likely to fall at between about 2 and 3 % annually. This figure reflects an average decline of around 5 % annually from post-peak fields being partially offset by increasing production from late fields coming on-stream. A 2–3 % decline in conventional oil represents an annual loss of global production of about 1.5–2.0 Mb/d; i.e., a decade's decline will give a loss of around 15 Mb/d. If the extra supply from the non-conventional oils and other liquids is not enough to offset this loss, and meet potential rising demand in addition, then global 'oil shocks' are inevitable unless demand for oil is curtailed by other factors. We look at forecasts of this supply in Chap. 4.

For a list of past and expected peak dates of production of 'all-liquids' by country from one company's forecast model, that of Globalshift Ltd., go to www.globalshift.co.uk.

2.6 The Second Half of the Oil Age

Colin Campbell, a noted analyst in the field, calls the new era of global oil supply that we are entering 'The Second half of the Oil Age'. Given that for the century and a half of the 'First half of the Oil Age' the world used primarily conventional oil from oil fields, what would indicate the start of the 'Second half'?

There are several candidates to mark this transition point. Arguably the 'Second half' starts at the point that we are facing now, where the world's ability to *increase* production of conventional oil becomes insufficient to meet its increasing demand for oil. It is true that currently some of the conventional oil that could come to market is being held back by some suppliers, but taking this constraint into account and also recognising that most suppliers of conventional oil are now past their 'mid-point peaks' in production, for some years now the world has required increasing production of the non-conventional oils (primarily natural gas liquids, tar sands oils, shale oil from fracking, and biofuels) to make up for what conventional oil in fields cannot supply. This in turn pushed up the price of oil since 2005 to levels that damaged global economic activity.

A later point for entry into the 'Second half of the Oil Age' is when the global production of conventional oil stops increasing and goes into decline, driven primarily by lack of recoverable resource of this type of oil. At the time of writing (Summer 2015) it is not clear whether this point has been passed—the IEA has perhaps suggested it was passed in 2006—or if it will be some time in the future; the date in part depends on how 'conventional oil' is defined, and also on the extent that oil producers find the application of enhanced oil recovery measures in conventional oil fields to be profitable. But when this point occurs, the increased production from the non-conventional oils must not only be enough to meet increased demand, but also to offset the decline in the production of conventional oil.

This leads to the final point in time where entry to the 'Second half of the Oil Age' becomes apparent. As mentioned, the expected fall in the global production of conventional oil, once past its 'resource-limited' peak, is likely to be of the order of 2–3 % per year. It is not clear that production of the non-conventional oils can take up the slack, nor at what price; and a number of forecasting models suggests that this may not be possible (Chap. 4). If this is indeed the case, i.e., if insufficient non-conventional oil production comes forward, then global oil production *in total* declines, and the 'Second half of the Oil Age' is well and truly here.

For forecasting oil production, so far in this book we have largely relied on the rule of thumb of 'peak at mid-point'. This is an extremely valuable and generally robust approach, but necessarily only a broad approximation. The next chapter gives a brief history of some of the more accurate approaches that have been used.

Chapter 3
A Brief History of Forecasting Peak Oil

Here we turn to the question of how to forecast peak oil. This is a large topic, and here only a few of the main approaches that have been used are highlighted. Additional detail on oil forecasting is given in Annex 4; and specific forecasts from a variety of individuals and organisations are discussed at greater length in Annex 5.

3.1 Forecasts for Conventional Oil Production

We start this chapter by focussing on *conventional* oil. A variety of methods exist to forecast the likely date of the peak conventional oil production in a region or globally, but the main issues to recognise are:

- Until the rate of conventional oil *discovery* in a region has peaked, forecasts of production are likely to be underestimates. This is because only when discovery has peaked that one can have a reasonable view of *future* discovery, and hence, and augmenting this with geology and other information, be able to have a reasonable view also of future production.
- As mentioned earlier, to know the rate of oil discovery, and hence know when discovery has peaked, proved reserves data cannot be used, and the oil industry proved-plus-probable (2P) data must be used instead.
- Once *discovery* has peaked and is in decline, 'field-aggregate' methods, such as mid-point peaking, can be used to forecast production.
- Where such methods rely on an estimate for the region's ultimately recoverable reserves (URR), this estimate needs to be realistic compared to the 2P discovery trend.
- Later still, once the yet-to-find is sufficiently small, and good field and basin knowledge for the region exists, by-field 'bottom-up' models can be reliably employed.

The next sections give a selective look at the some of the methods that have been used to forecast oil production over the years.

R.W. Bentley, *Introduction to Peak Oil*, Lecture Notes in Energy 34,
DOI 10.1007/978-3-319-26372-4_3

3.1.1 Oil Forecasts Prior to 1956

Peak oil sceptics point out the incorrect forecasts over many years for either the US or global peaks of conventional oil production. This view has some validity. Oil has long been a vital resource, and it is only natural that people have questioned its future availability on the basis of information available at the time.

Forecasts for the US made prior to the discovery of the East Texas field in the US in 1930, and the subsequent peak of US discoveries in new fields in the mid-1930s, were always likely to be underestimates. Likewise, forecasts for the world as a whole prior to the discovery of Ghawar in Saudi Arabia in 1948, and the subsequent peak of global discoveries of oil in new fields in the mid-1960s, were also likely to underestimate future production. However, once these *discovery* peaks had been passed, forecasting future production peaks for conventional oil was relatively straightforward, provided good data and correct approaches were used.

For example, when Hubbert made his 1956 prediction of the US Lower-48 conventional oil peak being between 1965 and 1970, discovery of US oil in new fields had been in decline for about 20 years. And for the world, the 'technically-based' forecasts from the mid-1970s to 1980s discussed below, of the global conventional oil peak around the year 2000, were made when global discovery of oil in new fields had been in decline for from one to two decades (see Figs. 2.21–2.23; though also see the discussion on 'reserves growth' in Annex 4).

Note that Hubbert's was not the only US or world forecast around these dates, a similar US forecast was made by Pogue and Hill a month earlier than Hubbert; and world oil forecasts made around that period are referenced in National Petroleum Council (2007).

3.1.2 Hubbert's Oil Forecasts

Before we discuss some of the later forecasts, we briefly look at those of Hubbert. He is often seen as the 'father' of peak oil forecasting, but his evolution as a forecaster carries important lessons for today.

In 1938, Hubbert made a forecast for the US peak that predicted peak too early by 20 years, writing that "*it seems doubtful that* [the beginning of the US oil decline] *can be postponed any later than 1950, and possibly not that long*" (Hubbert 1938). He based this on the fact that "*the easy discoveries have already been made*" (which was correct, as US 2P discovery peaked in the mid-1930s), and on the then-size of proved US reserves, at "*11 or 12 billions of barrels*". It was here his error lay, as he wrote that proved reserves represented "*the oil already discovered*". Like many to come after him, he had been led astray by thinking that proved (i.e. 1P) reserves were a useful measure of the amount of oil discovered. The actual US proved-plus-probable (2P) reserves at that date were certainly much greater; East Texas alone contained over 5 Gb. Today's backdated value for the size of US 2P reserves in 1938 is 95 Gb (see Fig. A2.2), i.e. about eight times the '11 or 12 Gb' that Hubbert quoted for the then-proved reserves!

By contrast, by 1956 for his well-known forecast of US Lower-48 peak, Hubbert had realised this mistake over proved reserves, and now relied not on these reserves, but instead used two oil industry estimates of the Lower-48 *original endowment* (URR) of conventional oil. As explained in Annex 5, these estimates were 150 Gb (Hubbert's preferred value) and 200 Gb (the latter from a Delphi enquiry by Wallace Pratt). Since the US had produced 52 Gb of oil by 1956 it was a simple matter to use a hand-drawn total production curve, based on roughly mid-point peaking, to show that the date of the US Lower-48 conventional oil peak could not be much different from 1965 or 1970 in the two cases (see Fig. 1 in Sorrell and Speirs 2014).

The crucial aspect of this forecast was that it did not correspond with most people's view that there would not be a US peak anytime soon. As Hubbert said, on the above data (a URR of either 150 or 200 Gb, and only 52 Gb used) "*the intuitive answer* [on peak] *was 'not in our lifetimes'. Our grandchildren may have to worry about it*".

Subsequent to the 1956 forecast, Hubbert developed a number of techniques to generate his own estimates for the US Lower-48 URR for conventional oil; see Annex 5 and Bentley and Boyle (2008). Most of these estimated this URR from around 170 Gb to under 200 Gb, so when the US peak for conventional oil (including condensate, but excluding NGPLs) occurred in 1970 at ~99 Gb cumulative produced, Hubbert's use of a reasonable URR plus mid-point peak was justified.

This dramatic reversal in US production (from going up to going down) was itself a major turning point in oil's history. Prior to this, OPEC's attempts to raise the oil price had failed in the face of the US opening its taps, see Yergin's *The Prize* (1991). But with the bulk of the World's oil then coming from only the US, the Soviet Union and the Middle East, the peak of US production in 1970 led to OPEC being able to control the price, and this allowed the 1973 and 1978 oil shocks to occur (see Bentley and Bentley 2015b).

The lessons for today from Hubbert's work are:

- Do not use 1P data. If 2P data are not available, generate these by applying an historical 'growth function' to the 1P data.
- Define the region being modelled (e.g. Lower-48) and the type of oil (e.g. conventional oil, i.e. flowable oil in fields); many subsequent analysts have criticised Hubbert's work based on not understanding these constraints.
- Use realistic (i.e. industry accepted) URR estimates, and predict peak production as (roughly) mid-point, or by other methods.
- Understand the errors of others, especially those with high URR estimates (e.g. Zapp, see Annex 5).

3.1.3 Other 'Technically-Based' Forecasts of Peak Conventional Oil Production, Made from the Early 1970s to Early 1980s

Next we consider a series of forecasts, in addition to Hubbert's, for the date of peak for global conventional oil production made between the early 1970s and early 1980s by a range of individuals and organisations. Over the decade or more following the oil shock of 1973 there was a major dichotomy in the perception of future global oil supply, and one which impacts the thinking of many oil analysts still to this day.

In much of the general culture at that time, including within government, by the media, and in view of most analysts, oil was seen as a scare resource, expected to become increasingly so, and which could well 'run out' in only 30 years or so. This thinking stemmed from the then-size of the global *proved* reserves (at 620 Gb in 1970; and 655 Gb in 1980) which represented about 30 years' of global production at the time; see for example President Carter's 'Moral Equivalent of War' speech, or the otherwise excellent study *Energy Future* by the Energy Project at the Harvard Business School (Stobaugh and Yergin 1979).[8]

In stark contrast, a number of analysts (mostly scientists) at that date recognised that then-current estimates for the global *URR* of conventional oil, at typically between 1800 and 2500 Gb, stood at a much higher level than the figure for just the proved reserves. Since only 240 Gb of this URR total had been produced by 1970 (and 460 Gb by 1980), it was relatively easy to use a 'mid-point' approach to forecast that the global production of conventional oil could continue to *increase* out to about the year 2000 (i.e. for about 20 or 30 years) before only then beginning to decline. See, for example, the presentation to ASPO 2003 conference by Andrews and Udall (Andrews and Udall 2015), or Bentley and Boyle (2008).

These 'technically-based' forecasts included:

- ESSO, 1972: Quoted a global conventional oil ultimate of 2100 Gb, and hence expected oil to become "*increasingly scarce from about the year 2000*". (The Ecologist 1972, pp 18 and 130.).
- B. Ward and R. Dubois, in a landmark environmental report to the United Nations (1972, p184): "*One of the most quoted estimates for usable reserves* [of oil] *is some 2500 billion barrels. This sounds very large, but the increase in demand foreseen over the next three decades makes it likely that peak production will have been reached by the year 2000. Thereafter it will decline*".
- The UK Department of Energy (1976, p12): Commenting on the expected date of the UK peak, noted that the world peak would not be far behind, at "*about* [the year] *2000.*"
- P. Ehrlich et al. (1977, pp 400–404): Predicted the global oil peak date as the year 2000, based on a 'high-estimate' for conventional oil endowment of 10,900 trillion MJ ($\sim$1900 Gb).

- Shell (1979, p1): Global oil production to "*plateau within the next 25 years.*" (But did not specify the data behind this forecast.)
- World Bank: Ultimate of 1900 Gb; oil expected to "*plateau around the turn of the century*" (World Bank 1981, p 37, 46.)

A number of other authorities gave estimates for the global oil ultimate, but did not carry these through to predictions for date of peak, for example:

- The Science Policy Research Unit at Sussex University, 1974: Range of World oil ultimate estimates: 1800–2480 Gb (Cole et al. 1974).
- The World Energy Conference (now the World Energy Council), ~1980: Referred to a study by the Institut Français du Pétrole estimating world oil ultimate as 1803 Gb. In commenting, J. Keily noted "*The world can have the energy it needs for the rest of the 20th century. But ... with a false sense of security, many will not look over the horizon to the early part of the 21st century. ... It is only by looking beyond the early 2000s that we can see how fast the change will come*" (Keily 1980, pp 26–32.).
- D. Meadows et al. in *Beyond the Limits* (sequel to *Limits to Growth*), 1992: Range for world oil ultimate: 1800–2500 Gb.

Note that the above estimates referred only to conventional oil; and almost certainly excluded NGLs. The possibilities of obtaining oil from tar sands, very heavy oil deposits, kerogen shales, and by conversion of coal or gas were all very well known at the time, but were not included in these estimates of URR as used to predict the date of peak.

3.1.4 Forecasts from Campbell/NPD, 1991; Petroconsultants (Laherrère, Demaison, Perrodon and Campbell), 1994–1996; and Campbell and Laherrère subsequently

Now we turn to an important more recent set of related forecasts of the global oil peak, those associated with the Petroconsultants consultancy in the mid-1990s.

(i) Oil forecasts from the Campbell/NPD study, 1991

We start with the study initiated by Colin Campbell in collaboration with the Norwegian Petroleum Directorate (NPD), and published as *The Golden Century of Oil 1950–2050* (Campbell 1991).

Campbell worked as an oil exploration geologist with Texaco, BP, Amoco, Shenandoah, Aran Energy and Fina; and studied, *inter alia*, the geology of Ireland, Borneo, the West Indies, Colombia, Australia, Papua, many further Latin American countries, Europe and the North Sea. He came to consider global oil depletion after being asked by BP in 1966 to write an appraisal of the oil prospectivity of Colombia, and later by Amoco for a similar study for Latin America as a whole. As he reports "a pattern began to emerge" of reduced opportunities to look for new oil. This led, when Executive Vice President of Fina's Norwegian operations, to his suggestion that the NPD be funded to study future global oil production, which resulted in *The Golden Century of Oil*.

This study looked at the production of conventional oil "*as produced in oilfields under conditions as we know them today*" (Campbell 1991, p 39), and did *not* include NGLs nor "*oil from tar or oilshales, nor with the proceeds of unconventional enhanced recovery techniques, although they are likely to become more important in the future*" (p xi). At that time Campbell was not aware of the necessity to use 2P data, and the study was based on the "*1989 Production and* [proved] *Reserve figures, mainly as reported by the Oil and Gas Journal*".

Three scenarios of global conventional oil production were produced, two with plateaux (from 1989 and 1992, respectively), and one with a sharp peak in production at about 1997. Campbell stressed the uncertainty of the data and of the modelling, and wrote that "*The scenarios, developed here, are therefore intended more to draw attention to the general limits of the underlying resource base than to claim to depict the precise shape of future production*".

Nevertheless, despite its reliance on 1P data, this was a very valuable study. At that time, the collapse of global oil prices from 1985, fed by rising production oil from Alaska, Mexico and the North Sea, had led to a dramatic switch in the views of most oil analysts, who now—contrary to their earlier warnings of ever-increasing oil shortage—became convinced that 'all past forecasts are wrong'; that OPEC's power, once so feared, was a phantom; and that investment could always turn 'resources into reserves'—as had apparently been so clearly demonstrated by growing production from the new provinces mentioned above (despite all of these being discovered prior to 1973). Thus the Campbell/NPD study provided evidence that this new widely accepted 'oil cornucopia' view was also wrong, and that the 'technically-based' forecasts of the 1970s and early 1980s listed above, with the global conventional peak expected around the year 2000, were still correct.

(ii) Petroconsultants studies and oil forecasts, 1994–1996

The next studies in this series were those by Petroconsultants between 1994 and 1996. At that time, and arguably still with its successor company IHS Energy, Petroconsultants held the world's best oil and gas dataset, which was contributed to and used by all the oil majors.[9]

In the early 1990s, Jean Laherrère was an exploration geologist, geophysicist and manager with Total with experience in Algeria, Australia and Canada. He had become interested in the use of logistic curves to model natural phenomena, including oil depletion, and suggested to Petroconsultants that they use this approach to model global supply. At around the same time Dr. George Leckie of that company, having seen Campbell's *The Golden Age of Oil*, suggested this study should be redone, but using 2P data.

These suggestions were taken forward, and a series of landmark consultancy reports were produced by Petroconsultants based on a 'petroleum system' approach (source rock, burial history, migration, reservoir and seal), Petroconsultants (1994–1996). URR estimates for conventional oil (ex NGLs) were made for a wide range of countries on the basis of a number of techniques, and also an estimate for the global conventional oil URR (ex NGLs) of 1800 Gb.

The results from these reports were summarised in a *Scientific American* article in March 1998: *The End of Cheap Oil* (Campbell and Laherrère 1998). This said "*Barring a global recession, it seems most likely that world production of conventional oil will peak during the first decade of the 21st century*". It continued "*Perhaps surprisingly, that prediction does not shift much even if our estimates are a few hundred billion barrels high or low*".[10]

The Petroconsultant studies on which this article was based also included examination of the large resources of non-conventional oil likely to be accessed post the conventional oil peak. The *Scientific American* article mentions Orinoco oil, Canadian and FSU tar sands, and shale deposits, but says "our … estimate is that only 700 Gb will be produced from unconventional reserves over the next 60 years". The study on the non-conventional hydrocarbons was published not by Petroconsultants, but by *The Petroleum Economist* (Perrodon et al. 1998).

Perhaps not surprisingly, and as Hubbert had found in 1956, predicting a fairly near date for the oil peak was not welcome everywhere. One oil company put pressure on Petroconsultants to suppress the report, and the American Petroleum Institute responded to it with a Discussion Paper: *Are We Running Out of Oil?* (Porter 1995). In addition, the *Scientific American* article based on the Petroconsultants studies was dismissed as 'fixed-resource nonsense' by a number of respected oil analysts (see Chap. 5 for some of the arguments used).

(iii) Subsequent oil forecasts by Campbell and Laherrère

Following the above consultancy studies, Campbell and Laherrère have continued to model global oil supply, and to publish their findings widely, see Annex 5, and a number of their charts are used in this book. For detailed descriptions of Campbell's and Laherrère's current oil forecasting models, see Campbell (2015), and Laherrère (2015).

3.1.5 *Forecasts Compared in UKERC* Global Oil Depletion *Report, 2009*

Next we look at a fairly comprehensive survey of oil forecasts carried out in 2009 for the UK Energy Research Centre, as part of the report *Global Oil Depletion* (Sorrell et al. 2009). This study set out to answer the question: '*What evidence is there to support the proposition that the global supply of 'conventional oil' will be constrained by physical depletion before 2030?*' Many of the forecasts, however, also covered all-oil or all-liquids.

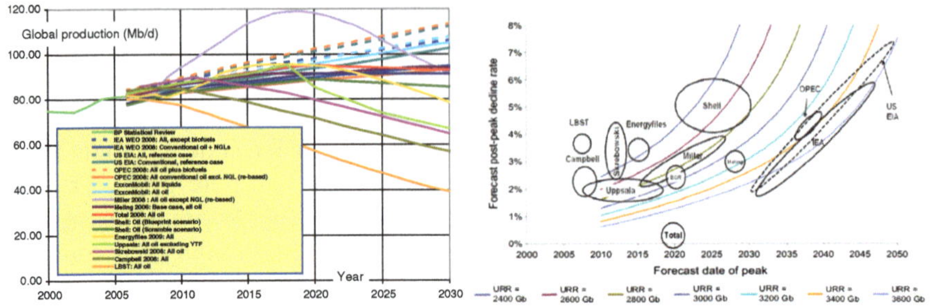

Fig. 3.1 Forecasts of global oil production from the UKERC Global Oil Depletion report (2009). *Left chart* Forecast global oil production (covering various categories of oil) versus date.

Legend (from *top*) is:
- BP *Statistical Review* (historical data);
- IEA WEO 2008: All-oil, except biofuels;
  Ditto: Conventional oil + NGLs;
- US EIA: All-oil, reference case;
  Ditto: Conventional oil, reference case;
- OPEC 2008: All-oil plus biofuels;
  Ditto: All conventional oil excl. NGL (re-based);
- ExxonMobil: All-liquids; ExxonMobil: All-oil;
- Miller 2008: All-oil except NGL (re-based);
- Meling 2006: Base case, all-oil; Total 2008: All–oil;
- Shell: Oil (Blueprint scenario);
  Ditto: (Scramble scenario);
- Energyfiles 2009: All-oil;
- Uppsala University: All-oil except yet-to-find;
- Chris Skrebowski 2008: All-oil;
- Campbell 2008: All-oil;
- Ludwig-Bölkow-Systemtechnik GmbH (LBST): All-oil.

Right chart Date of peak versus post-peak decline rate and size of ultimately recoverable resource (URR). Graph developed by Dr. Richard Miller to examine the differences between the forecasts for (primarily) conventional oil. Isolines represent the assumed or implied global URR. Rate of production increase prior to the peak is set to 1.3 % per year. Mapping of individual forecasts onto the graph involves some judgment. Conventional oil here includes crude oil, condensate and NGLs, but in some cases also included production from currently operating and planned oil sands production as this was difficult to separate out. Excluded is oil from oil sands plants not yet planned, Light tight oil, oil from kerogen shale, and other liquids (GTLs, CTLs and biofuels). Note that Total includes extra heavy oil in its model

Figure 3.1 (left) shows a summary of forecasts in the report. At that date there were three distinct classes of forecast:

- Forecasts for a roughly linear increasing trajectory for global oil production out to 2030 (the end of the forecast period). These forecasts were mainly for all-oil from the 'mainstream' forecasters: the IEA, US EIA, OPEC, and ExxonMobil.
- Forecasts that saw production increasing out to 2030, but flattening out. These were either from the 'mainstream' forecasters but for conventional oil, or were

for 'all-oil' from the other oil companies reviewed: Shell, Total and StatoilHydro (where the latter was a private forecast from a senior employee).
- Forecasts that showed global oil production peaking before 2030. These were mainly from the 'independent' forecasters: Energyfiles, LBST, University of Uppsala, Miller, Campbell and Skrebowski.

The difference between the forecasts examined was not explained simply by the categories of oil covered. For example, the US EIA had a roughly linear plot for conventional oil production to reach ~103 Mb/d by 2030, whereas at the other end of the spectrum LBST forecast 'all-oil' production to fall to only 40 Mb/d by the same date. The main difference between the forecasts was that between the 'mainstream' oil forecasters and the independents. (And note that only a few years earlier the 'mainstream' forecasts had been for global supply to grow much faster still, to reach around 120 Mb/d by 2020.)

To examine the assumptions—not always explicit—in the forecasts for *conventional* oil, and to see how likely these assumptions were, Dr. Richard Miller devised the plot of Fig. 3.1 (right). On such a plot forecasts that show global production peaking before 2030 are to the left, while the 'quasi-linear' forecasts are to the right. As the plot shows, most forecasts which had the date of peak as 2020 or before (LBST, Campbell, Skrebowski, Energyfiles, Uppsala University, BGR and Miller) typically had fairly low URR's and also 'medium' post-peak decline rates (from 2 to 4 % p.a.), while the 'mainstream' forecasts of OPEC, US EIA and the IEA were constrained to have high URRs for conventional oil (from ~3250 Gb to ~3800 Gb); and also post-peak decline rates that were quite high (~>5 % p.a.) if the date for peak was assumed to be after 2040.

Thus—in broad terms—the large difference between the forecasts in Fig. 3.1 for global conventional oil production could be explained mainly by differences in the URRs assumed or implied for conventional oil; and in the post-peak decline rates assumed or implied. The question for forecasters therefore is what are the most realistic values for these variables? This, with related topics, is discussed next.

3.1.6 Current Forecasts for Conventional Oil

In looking at current forecasts, it is necessary to differentiate between two classes of forecast: 'field-aggregate' and 'bottom-up by field' (sometimes abbreviated here to 'aggregate' and 'bottom-up', though elsewhere these terms can mean other modelling approaches).

In a field-aggregate forecast there is no attempt to model individual field production profiles, but instead to use one or more general functions that are known or assumed to reflect the behaviour of large groups of fields.

Probably, the best known of these is to model all the world's fields as a single group, and apply a Hubbert curve to predict total production. Alternatively, more granular approaches can be used. Campbell, for example, for 'Regular conventional' oil models the combined production of all fields in a region (usually a country), and uses different approaches depending on whether the country is before or after its peak in production (the latter usually being the case); and also if it is a Middle East Gulf country.

A different field-aggregate approach forecasts production by summing a sequence of logistic curves, where each represents a different 'phase' of oil production (such as from a later exploration phase in a basin), and also different categories of oil, such as tar sands and other non-conventionals, as in Laherrère's global model. Other aggregate approaches exist, see Annex 4.

By contrast, a bottom-up by-field forecast predicts the production of *individual fields*, and sums these outputs to give total production in a region, and, if regions are summed, the global output.

In practice it is not possible to model all fields, there are just too many (for the world, over 30,000 outside the US); and often data for large numbers of individual fields are not available. In such cases, the bottom-up forecast aggregates the smaller fields, and also unquantified fields, into groups, and models them either as a large single field, or in some other manner. Quantities of oil yet-to-find are usually modelled by a sequence of fictitious fields, following some size function, coming-on-stream at a defined rate, while technology gain and other sources of oil (such as the non-conventionals) are handled by other means. The key feature of bottom-up by-field modelling is that the bulk of the forecast volume is derived from assumed production profiles of individual fields, even when a large amount of combining of fields takes place.

Both methods (aggregate, and bottom-up) have their strengths and weaknesses. While yet-to-find in a region is large, applying a bottom-up model is difficult. But now that most regions in the world have comparatively small yet-to-finds of conventional oil compared to their URRs, bottom-up models are fully applicable. Such models require a great deal of data and assumptions, and therefore may be hard to check by others, and can lead to forecasts covering a wide range of outcomes if all sensible variations in these parameters are assessed. Aggregate models, by contrast, capture less of the detail of the 'real world', but have the value that their assumptions are usually simple and robust. In predicting the increasing global oil supply constraints (and hence high prices) that occurred since about 2004, those analysts using aggregate models have generally made the best predictions. But going forward both types of model are needed to properly understand the oil future.

There is a wide range of data and modelling issues that need consideration when creating (or critiquing) current oil forecasts, and some of the main ones are discussed in Annex 4. They include getting reliable data, modelling different categories of oil, allowing for 'reserves growth', treatment of 'fallow fields', use of realistic decline and depletion rates, application of a 'Hubbert curve' to a region or the world, proper use of URR estimates, aggregate models that do not assume an URR and 'capacity' versus 'realistic' by-field forecasts. As mentioned above, probably the two most important of these issues are accessing reliable data and the use of an estimate for URR that is realistic in terms the regional or global discovery history.

3.2 Forecasting Production of Non-conventional Oils, & Other Liquids

Finally, in this chapter we turn to the important issue of forecasting production of non-conventional oils, and of the 'other liquids'.

Note first that for non-conventional oils, and also the other liquids such as gas or coal to liquids, and biofuels, in the main none of the concepts of field, field size distribution (and hence discovery effort) and field decline are relevant, and hence neither is the concept of 'mid-point' peaking for production of these liquids. So while production of a non-conventional oil or other liquid may indeed peak, it will be from different causes than for conventional oil.

For these non-conventional oils and other liquids, mostly the potential resources are very large, and for almost all, their locations are already known so exploration costs are low. (The exception on resource size is biofuels, because of competing requirements for food and biodiversity.)

However, there is a range of constraints likely to limit production of these fuels. These include

- technological readiness;
- a high investment requirement, typically, per barrel produced;
- other resource requirements (energy, water, land for waste);
- net-energy related cost and energy yield factors;
- CO_2 emissions.

Note that the availability of investment for non-conventional fuels is not straightforward. When the general energy price raises, the cost of all fuels, including the alternatives, also tends to rise because of the energy embodied in bringing these fuels on-stream. This effect was demonstrated for tar sand extraction costs during the 1970s oil shocks, for oil from shale (kerogen oil) as reported in Stobaugh and Yergin (1979, p 43), and more recently across a range of alternative fuels. The concept was also a key factor in the original *Limits to Growth* modelling. And if the energy price becomes also more volatile, then investment in alternative fuels is further discouraged, even during periods of lower price.

In the following, we look at three aspects of non-conventional oil supply of current focus: the availability of light tight oil; the decline in energy return on energy invested (EROEI); and CO_2 emissions, and hence the concept of 'stranded' fossil fuel assets.

Table 3.1 Global technically recoverable oil and gas *remaining* (where 'shale' here refers to 'light-tight' produced by fracking, and excludes some regions and categories of tight oil; and also oil from oil shale, i.e. from kerogen in source rocks following retorting). *Source* EIA (2013)

	Conventional	Shale ('tight')	Total
Oil (Gb)	3012	345	3357
Gas (Tcf)	15,583	7299	22,882
Gas (*Gboe*)	*3000*	*1400*	*4300*

The above data, in 'years of global supply at current use rates' translate to:

	Conventional	Shale ('tight')	Total
Oil ('years of supply')	95	10	~100 yrs
Gas ('years of supply')	130	60	~200 yrs

(a) Availability of light tight (shale) oil

The future production of light tight oil by hydraulic fracturing ('fracking') is usually discussed (at least in Europe) in terms of public acceptability relating to water pollution, seismic shocks and greenhouse gas emissions. But generally overlooked is the fact that the *total amount* of light tight oil (as opposed to tight gas) assessed as globally recoverable is not very large.

Figure 2.24 showed the IEA's estimate of ~240 Gb recoverable, while the data (Table 3.1) from the US EIA (2013) give a somewhat higher estimate, of 345 Gb.

Thus, if a 'mid-point' rule is applied to the EIA data, light tight oil can only shift the date of the global 'all-oil' peak by about 5 years, whereas shale gas might potentially shift the 'all-gas' peak by 30 years. While such a calculation is only indicative ('mid-point' not being applicable to non-conventional oils, and 'years of supply' being usually misleading!), it does highlight the relatively small quantity of this type of oil.

(b) Energy return on energy invested

Energy return on energy invested ('EROEI', or 'EROI') is a significant issue, but one not receiving attention in nearly all global oil models (and not in the main-stream 'all-energy' models, for that matter). The data are probably still not adequate to give a comprehensive picture, but none the less are sobering, as shown in Table 3.2.

The recently rapidly falling EROI ratio for conventional oil, if confirmed by other studies, is of concern as independently Hall and co-workers have suggested that modern society needs a minimum EROI ratio of perhaps 10–15:1 from its fuel sources to function in its current form (e.g. Hall 2008; Lambert et al. 2014). And even where EROI ratios are higher than this, being lower than in the past reduces society's overall wealth if not compensated for by productivity or efficiency gains elsewhere.

Recently, Campbell (2015) has incorporated EROI ratios into his global oil forecast, multiplying his forecast production levels for the various categories of oil

Table 3.2 Approximate ranges for EROI ratios

Conventional oil	Date: 1920	1970	1995	2006	2012
US	20	30	35	20	10
Global publicly traded			36	18	
Other energies					
Tar sands	1.5–8				
Coal	40–80				
Nuclear fission	4–16			(N.B. Old data, possibly out of date)	
Wind	10–28				
PV	2.5–8				
Biodiesel, gasohol	~3				

Sources Hall and Day (2009), Gagnon et al. (2009), Guilford et al. (2011), Hall et al. (2014), Hall, private communication. Data subject to revision

('Regular conventional', deepwater, tar sands, etc.) by the corresponding 'net yield' ratio to turn gross barrels of production by category of oil into a corresponding forecast of 'net-energy' barrels. Because the non-conventional oils have—in general—lower EROI ratios than conventional oil, the overall decline in global production is steeper in net-barrels terms than in gross barrels.

And in terms of investment required to produce future oil, as mentioned earlier note that the need for increasing quantities of a resource to extract that resource was one of the drivers of system collapse in the original *Limits to Growth* modelling (the other driver being rising costs from increasing pollution); see Meadows et al. (1972), and also (1992, 2004), and Bardi (2011, 2014).

Finally, an often overlooked aspect of EROI data is net-energy *rate* limits. These are limits to the maximum *rate* that an energy-producing technology (in this case, a new fuel source) can be usefully introduced, and reflect the fact that if the technology is introduced faster than the embodied energy required for new plant, the overall net-energy yield *during the growth phase* is negative. (For example photovoltaics, with about 200 GWp installed, have to date yielded no net-energy to mankind. This is partly due to having a moderate EROI ratio, but mainly to their rapid uptake, see, e.g., Dale and Benson 2013.) And even where technology is introduced at a slower pace, the net-energy yield can be significantly less than the energy yield as usually calculated. Most current global energy models do not take into account either EROI ratios or net-energy rate limits.

(c) CO_2 emissions

With IPCC-5 now issued, it has become even clearer that global CO_2 emissions will need to be controlled if dangerous climate change is to be avoided. This has impacts on the extent that conventional oil can be replaced by the non-conventionals, and hence the extent that some of the latter may have to remain in the ground as 'stranded assets'. The link between peak oil and climate change is discussed in Annex 6.

Chapter 4
When Should We Expect the Peak?

4.1 Peak in Fields and Regions

This chapter draws together the above information to answer the question: When should we expect the '*resource-limited*' production peaks, for the various categories of oil, in fields, regions, and the world?

For *fields*, see data on production from field operators, or sometimes relevant governments. However, for *forecast* peak dates for individual fields there may be no general sources, and owners of bottom-up models, who have such forecasts, need to be contacted.

For *regions*, and specifically for countries, there are good sources of forecasts. Campbell's *Atlas* (2013) should be consulted for 'Regular conventional' oil, with its essential plots of how much has been produced versus how much has been discovered, and hence forecasts for production. Rystad Energy's UCubeFree has plots by country for past and future oil production (but not of discovery), while Globalshift Ltd. has excellent plots on its website of past and future production by country for 'all-liquids' split by category (though again, not of discovery), and also a table of when countries will pass their oil peak, or have already done so. Other forecast owners, such as IHS CERA, can also be contacted for their information.

Forecasts by country are of particular interest to exporter countries that are still pre-peak. Nigeria, for example, would be wise to build government policies now to anticipate the significant changes—in food supply, energy subsidies, and industrial and commercial activity—that will need to be faced once the energy and revenue impacts of its oil production peak (due around 2020) become apparent.

4.2 The Global Peak

It is, however, the peak of *world oil production* that is of the greatest interest.

R.W. Bentley, *Introduction to Peak Oil*, Lecture Notes in Energy 34,
DOI 10.1007/978-3-319-26372-4_4

We start with a recent forecast of global 'all-liquids' production to 2030 by IHS CERA (Fig. 4.1, left graph) because this forecast is seen by some observers as being at the more optimistic end of the current spectrum. As the figure shows, and also Fig. 4.1 (right), this forecast sees no early peak in conventional oil, but instead an undulating plateau at around 85 Mb/d, starting from about 2020. For all-liquids CERA indicates a somewhat similar plateau, but at around 115 Mb/d, and starting about 2040 (beyond the end of the forecast period).

An earlier chart from CERA helps clarify the probable assumptions. That chart suggested that the 'Peak Oil' line in Fig. 4.1 (right) corresponds to a Campbell estimate for global conventional oil URR of 1920 Gb, whereas the CERA forecast at that date had a URR for conventional oil about 1000 Gb larger (at 2930 Gb); and for all-oil larger still by ~700 Gb (at 3610 Gb). Note that the Campbell and CERA URRs for conventional oil will differ somewhat in categories of oil included, in particular on NGLs (Note also that Campbell's 2014 estimate for the global 'all-conventional' oil URR is 2050 Gb.)

The two undulating plateaux shown in the IHS CERA forecast are far from what most economies are banking on. As Fig. 4.1 (left) indicates, even all-liquids cannot meet a significant increase in demand after about 2020. This is a serious situation for a world accustomed to a century and a half of ever-increasing quantities of oil to drive its economies.

Next, we compare the above IHS CERA forecast with one at the more cautious end of the 'forecast spectrum', that of Laherrère (2015).

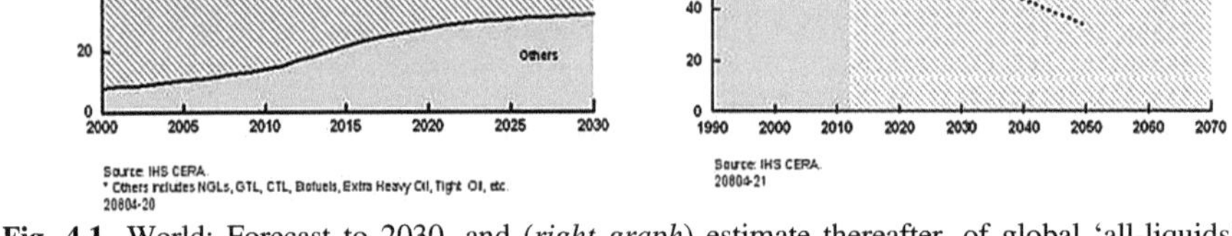

Fig. 4.1 World: Forecast to 2030, and (*right graph*) estimate thereafter, of global 'all-liquids' production to 2070, from IHS CERA.

Left Forecast to 2030. Shows: (**a**) production of conventional oil from fields currently in production (FIP), fields under development (FUD), fields under appraisal (FUA), and fields yet to find (YTF); and (**b**) of other liquids as 'Others', being that part of all-liquids that is not conventional oil, and includes "NGLs, GTL, CTL, Biofuels, Extra heavy oil, Tight oil, etc".

Right Indicative all-liquids production 2030–2070. 'Unconventionals' is the same as 'Others' in the *left plot*. Shows an approximate peak of all-liquids about 2040, in contrast to the 'Ever Upward' arrow that is currently indicative of 'mainstream' oil forecasts, such as those from the IEA or US EIA.

Source Jackson and Smith (2014)

Figure 4.2 shows Laherrère's current forecast for global all-liquids, and for its main components. Data are from a variety of industry sources, and as mentioned earlier, Laherrère adjusts industry discovery data to allow for Middle East, FSU and Orinoco overstatements (For the backdated 2P discovery data that Laherrère factors-in when generating this forecast, see Fig. 2.22).

The forecast of Fig. 4.2 gives the global all-liquids peak as around 2016. Laherrère notes that his forecast:

> *is modelled using several ultimates estimated from extrapolation of cumulative discoveries (creaming curve) or cumulative production. The first ultimate is for 'crude oil less extra heavy' at 2200 Gb, giving a peak in 2010. The second is for extra-heavy [oil] at 500 Gb, giving a peak in 2070. The third for NGLs at 300 Gb, is for a peak in 2025. A fourth ultimate refers to refinery gain of 70 Gb (3 % crude - XH).*

A fifth and final contribution in his model covers 'other liquids' (mainly biofuels). Here an 'ultimate' is not appropriate (being renewable), but instead an asymptote is currently set at 3 Mb/d (down from an assumed 5 Mb/d in an earlier version of the model). These assumptions taken together generate a global all-liquids peak around 2016, at around 95 Mb/d.

Figures 4.1 and 4.2 thus give two very contrasting forecasts: Laherrère's forecast is for the global all-liquids peak to occur about now, at ~95 Mb/d, while the IHS CERA forecast has instead, not a peak, but a global all-liquids plateau (at ~115 Mb/d) starting only around 2040.

How can we reconcile such different views? This can be done by comparing the data behind these two forecasts, starting by looking at the IHS Energy end-2011

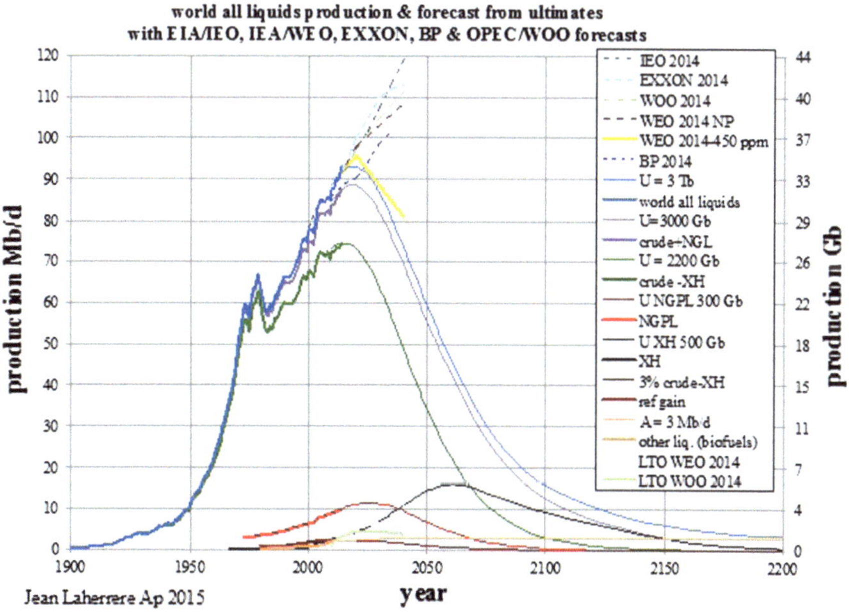

◀ **Fig. 4.2** World: Forecast of global 'all-liquids' production from Jean Laherrère to 2200, and showing the components of this. Also shown are forecasts from some of the 'mainstream' oil organisations, for comparison.

- Production of global all-liquids, 1900–2014;
- Forecast of global all-liquids supply 2015–2200, generated by the model described in Laherrère (2015), and based on URR values by category of liquid;
- Global all-liquids production forecasts generated by the US EIA, Exxon, OPEC, IEA and BP; and also IEA and OPEC forecasts for the production of 'light-tight' ('shale') oil.

Legend

- IEO 2014: US EIA *International Energy Outlook* all-liquids forecast, 2014;
- EXXON 2014: Exxon all-liquids forecast, 2014;
- WOO 2014: OPEC *World Oil Outlook* all-liquids forecast, 2014;
- WEO 2014 NP: IEA *World Energy Outlook*, New Policies scenario all-liquids forecast, 2014;
- WEO 2014-450 ppm: IEA *World Energy Outlook*, Scenario to meet 450 ppm CO2, all-liquids forecast, 2014;
- BP 2014: BP *Energy Outlook* all-liquids forecast, 2014;
- U = 3 Tb: 'Hubbert curve' (logistic derivative curve) for a URR of 3 Tb plus addition of refinery gain plus production of 'other liquids' (modelled here as primarily biofuels, but conceptually including small amounts of oil from kerogen, GTLs and CTLs). This curve is drawn so that its upside roughly matches past actual all-liquids production, and its total area = 3 Tb plus the addition of refinery gain and other liquids. The figure of 3 Tb itself results from summing a URR of 2200 Gb for 'conventional' oil (crude oil production including condensate but less the extra-heavies, and less NGPLs); plus a URR of 300 Gb for NGPLs; plus a URR of 500 Gb for the extra-heavy oils (primarily tar sands and Orinoco oil);
- world all-liquids: World all-liquids actual production;
- U = 3000 Gb: 'Hubbert curve' for a URR of 3000 Gb (=3 Tb) drawn so that its upside roughly matches past actual all-liquids production excluding refinery gain and 'other liquids', and its total area = 3 Tb;
- crude+NGL: Actual production of global all crude oil (comprising conventional oil, condensate, heavy oils, tar sands and Orinoco oil and 'light-tight' oil), plus all NGLs;
- U NPGL 300 Gb: 'Hubbert curve' for a URR of 300 Gb drawn so that it matches past actual natural gas plant liquids (NPGL) production data, and its total area = 300 Gb;
- NGPL: Actual production of NPGLs;
- U XH 500 Gb: 'Hubbert curve' for a URR of 500 Gb drawn so that it roughly matches past actual extra-heavy oil production data (primarily tar sands and Orinoco oils), and its total area = 500 Gb; 3 % crude—XH: A simulacrum of refinery gain, calculated as 3 % of total crude production less the extra-heavies production;
  ref gain: Actual refinery gain data;
- A = 3 Mb/d: Modelled production of 'other liquids', i.e. liquids for fuels not included in the categories modelled above, and consisting primarily of biofuels, but also of assumed small contributions from GTLs, CTLs, synthetic oils and oil retorted from kerogen either at the surface or in situ. Production is modelled as a curve reaching 3 Mb/d by ~2030, and holding constant thereafter;
- other liq. (biofuels): Actual production data for 'other liquids', primarily biofuels;
- LTO WEO 2014: The IEA *World Energy Outlook* 2014 forecast for production of 'light-tight' ('shale') oil;
- LTO WOO 2014: The OPEC *World Oil Outlook* 2014 forecast for production of 'light-tight' oil

data given in the text of Miller and Sorrell (2014), and plotted above in Fig. 2.23. These data give the global cumulative 2P discovery, for "crude oil, condensate, NGL, LPG, heavy oil and syncrude" at that date as ~2486 Gb, and the corresponding global cumulative production of this oil as 1248 Gb.[11]

From Fig. 2.23 one can estimate a 'reasonable medium-term extrapolation of 2P discovery' URR for these categories of oil, with this lying between perhaps ~2600 Gb and ~2800 Gb, i.e. well below the ~3600 Gb URR assumed for global all-oil in the earlier CERA forecast mentioned above. If NGLs, tar sands, other very heavy oils and light-tight oil are then taken out, this puts this current global 'extrapolation' URR for *conventional* oil at about from 2300 to 2500 Gb, depending on assumptions. This range agrees pretty well with the much earlier estimates given above from the early 1970s and 1980s for the global URR for conventional oil of between 1800 and 2500 Gb.

If then adjusting down further for the possible overstatement of 2P reserves for OPEC and FSU countries, and for some of the oil in Venezuela, the global conventional oil URR falls to about 2000–2200 Gb, which agrees with Laherrère's current 'extrapolation' URR shown in Fig. 2.22, of 2200 Gb for crude oil less NGLs and less extra-heavies.

Thus the difference between these two forecasts lies mainly in assumptions about the data, with Laherrère taking a considerably more conservative view of the quantities of oil likely to be available than does IHS CERA and with Laherrère's URR matching more closely the 'extrapolation of 2P discovery' data.

To broaden this view on when to expect the global peak for different categories of oil first we look also at recent forecasts from Campbell (Fig. A4.2 right, in Annex 4), and from Globalshift Ltd. and Rystad Energy, Fig. 4.3.

Comparing these three forecasts we have:

- Campbell forecasts the peak of global 'Regular conventional' oil production at about 2005; and a bumpy plateau for all-oil production from about 2004 to 2022, with the actual peak around 2015 at about 85 Mb/d. (Note that Campbell does not model contributions from GTLs or CTLs, assuming these to be small over any reasonable timeframe; nor from biofuel as not being a fossil resource.) Incidentally, for global '*all-hydrocarbons*' (oil plus gas), but excluding the oil categories just mentioned and also possible gas contributions from UCG and methane hydrates, Campbell forecasts the date of the global peak as soon, at around 2020, at a total production of ~59 Gboe/yr.
- Globalshift Ltd. forecasts global all-liquids (including biofuels) production to peak at about 2025, or a little after. For conventional onshore oil the peak is shown as a bit after 2020, possibly in part reflecting the fallow fields still awaiting development.

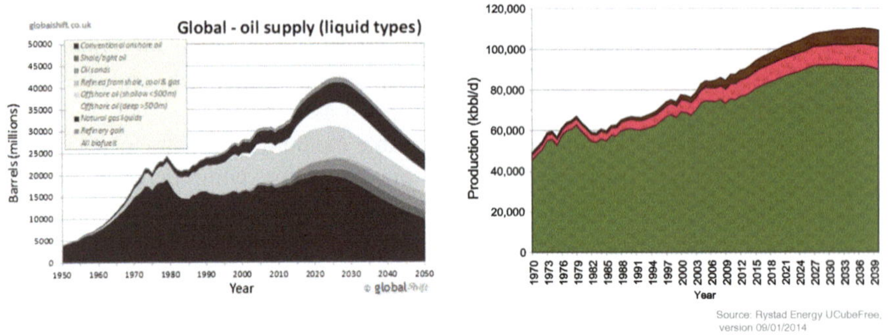

Fig. 4.3 World: Forecasts from Globalshift Ltd. and Rystad Energy.

Left chart Globalshift Ltd. plot of all-liquids; actuals: 1950–2012; forecast: 2013–2050. Category (from *bottom* of *chart*) Conventional onshore oil; Shale/tight oil; Oil sands; Refined from shale, coal and gas; Offshore oil (shallow <500 m); Offshore oil (deep >500 m); Natural gas liquids; Refinery gain; All biofuels. *Vertical axis* Mb/yr.

Source Globalshift Ltd., March 2014. For a description of this forecast model see Smith (2015).

Right chart Rystad Energy plot of all-liquids; actuals: 1970–2012; forecast: 2013–2040. Category (from *bottom* of *chart*) Conventional and non-conventional oil; NGLs; Other liquids'. *Vertical axis* Mb/day in 20,000 Mb/d increments to 120,000 Mb/d.

Source Rystad Energy: UCubeFree licence, February 2014

– Rystad Energy is more optimistic for global all-liquids production, forecasting near-plateau from about 2027, but with the actual peak about a decade later, at around 2037. For details of Rystad Energy's detailed asset-based price-driven model see Wold (2015).

We end this examination of current forecasts by looking at three from the 'mainstream' oil forecasting organisations, Figs. 4.4, 4.5 and 4.6.

The IEA's 2011 forecast puts the global production of conventional oil (here classed as 'Crude oil') as staying flat to 2035, with the increase needed to meet demand over this period coming from NGLs, non-conventional oils, refinery gain and biofuels.

This figure from BP shows a very similar forecast to the IEA's: global conventional oil production remains flat throughout the forecast horizon.

In the figure from ExxonMobil, as with Figs. 4.4 and 4.5, the global production of conventional oil is forecast to remain flat (indeed, slightly decreasing) throughout the forecast horizon, here to 2040.

Though nearly all the 'mainstream' forecasters, such as the IEA, US EIA, OPEC and ExxonMobil, have a very poor record of past forecasts (see Annex 5), some—as these three cases show—have begun to see at least a plateau for conventional oil production going forward, while for 'all-liquids', though none yet see a production peak within their forecast horizons, all see production increases as tapering off.

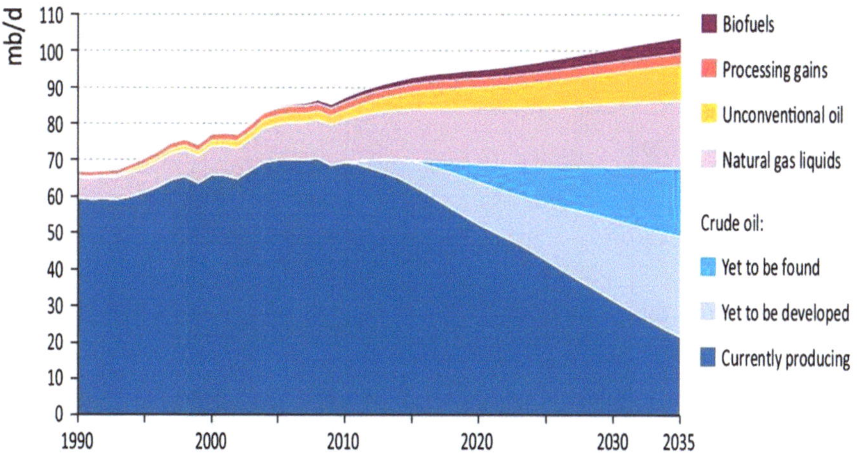

Fig. 4.4 IEA *WEO* 2011 forecast of Global 'All-liquids' Production to 2035. *Source* IEA *World Energy Outlook*, 2011

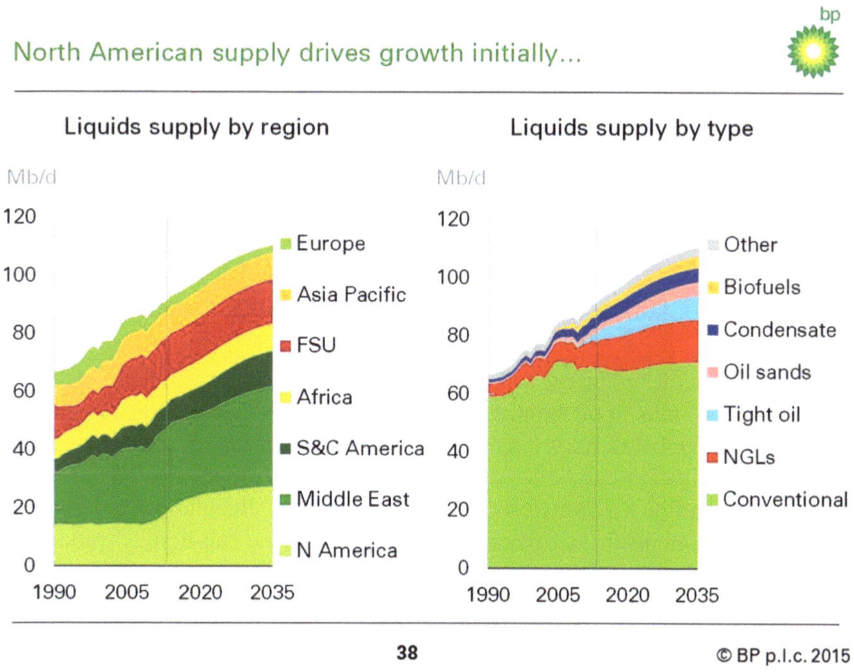

Fig. 4.5 BP's 2015 *Energy Outlook 2035* Forecast of Global 'All-liquids' production to 2035. *Source* BP website, accessed 20 May 2015, *Energy Outlook 2035*

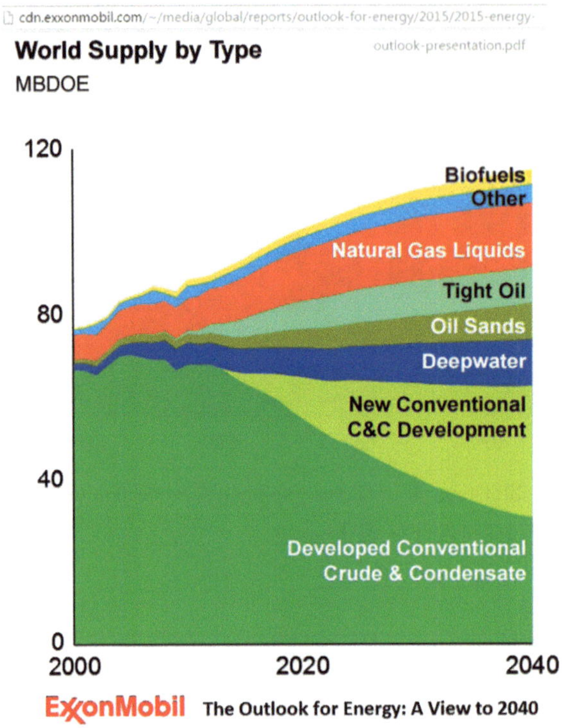

Fig. 4.6 ExxonMobil Forecast to 2040 of World Liquids supply, by type of liquid. *Source* Data from the ExxonMobil website; supplied to the author by M. Mushalik (http://crudeoilpeak.info)

4.3 Current Data on Production by Category of Oil

Finally, in addition to current forecasts, we can gather at least indicative data on possible global limits to oil supply from recent data on oil production by category, Figs. 4.7, 4.8 and 4.9(a). (Note: As will be made clear in Chap. 5, caution is needed as current information can be a very misleading guide to future production.)

We start with Fig. 4.7, which shows the production of 'Regular conventional' oil as defined by Campbell, and reconciles this to the global 'all-liquids' production data.

As Fig. 4.7 indicates, the production of 'Regular conventional' oil has been in clear decline from 2005, with total conventional oil showing a slight decline since that date. Increases have come from tar sands, NGLs and biofuels, though these increases have been modest. Next we look at a different chart on global 'crude plus condensate' production, Fig. 4.8.

Figure 4.8 shows a similar picture to Fig. 4.7. Note that the data in Fig. 4.8 predate the small 2015 OPEC conventional oil production increases, which together

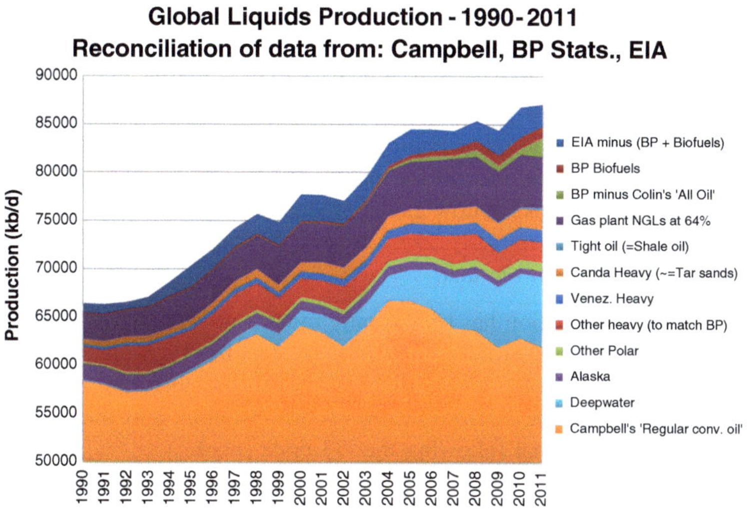

Fig. 4.7 Recent information on global oil production by category of oil.

Left chart Reconciling EIA 'total liquids' production data back to Campbell's 'Regular conventional' oil (see definition in Annex 1), 1990–2011.

Data sources Campbell, BP, US EIA

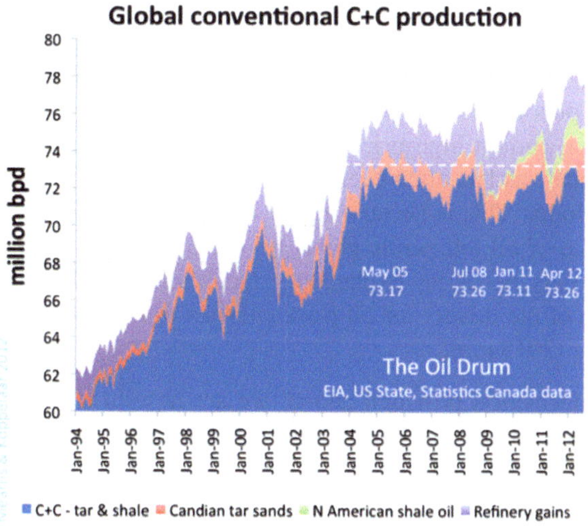

Fig. 4.8 Recent information on global oil production by category of oil. Components of US EIA 'All-oil' production data, 1994–2012.

Note 'C+C': Crude oil plus condensate.

Source The Oil Drum website; data sources as listed

with US non-conventional (light-tight) production and an only modest increase in demand, have helped drive the oil price back down to levels last seen in 2009.

Figure 4.9(a) looks in more detail at the components of all-oil production.

Figure 4.9(a) indicates that except for NGLs, the production to-date of the non-conventional oils, and other liquids, has been quite modest; conventional oil has supplied the bulk of global oil for many years.

Finally, Fig. 4.9(b) looks at how global production, and real-terms oil price have changed since 1965.

In Fig. 4.9(b), one can see that:

- Global all-oil production has shown only modest growth since 2005.
- The real-terms oil price rose from 2002 to that of the 1978 oil shock by 2008, fell in 2009 to a bit above that of the 1973 oil shock, then rose again. The fall since 2011, now (Sept. 2015) down to about ~$45/bbl, puts the real-terms oil price back to about that following the '73 price shock.
- For over four decades prior to 1973, the real-terms price of oil had averaged about $15/bbl in today's money, and global production had grown rapidly, at up to ~8 % a year. Once the shocks of the 1970s were passed, the price collapsed, but only down to about twice the pre-1973 level, averaging in real-terms ~ $30/bbl over the next 15 years or so (1985–2000), partly due to OPEC quotas on volumes, but probably primarily due to the marginal barrels becoming more expensive, with these coming from the North Sea, Alaska, offshore Africa and elsewhere. In this period production growth was considerably less, averaging somewhat under 2 % a year.
- More recently, with the oil price varying between $50/bbl and $100/bbl, production growth (of all-liquids) has been lower still, below 1 % a year, where now the marginal barrels are the additional conventional oil brought on by the high price, plus increasing production of the non-conventionals.
- Note that the current high price cannot be driven *fundamentally* by growth in demand alone—as has often been claimed, citing, e.g. rapidly rising demand in China and India. This is because in the over a century from 1861 to 1973, the average price of oil fell, while global oil demand for oil grew rapidly. Today's high price thus needs an explanatory factor in addition to demand. For an explanation of the drivers for oil price changes from 1861 to 2014, see Bentley and Bentley (2015a, b).

With the information provided above, we are now in a position to draw conclusions on probable dates of peak oil production, by category of oil.

4.4 Conclusions on Expected Dates of Peak

In the above sections we have looked at fairly recent global oil forecasts from IHS CERA, Laherrère, Campbell, Globalshift Ltd., Rystad Energy, the IEA, BP and ExxonMobil; and also at recent production data by category of oil.

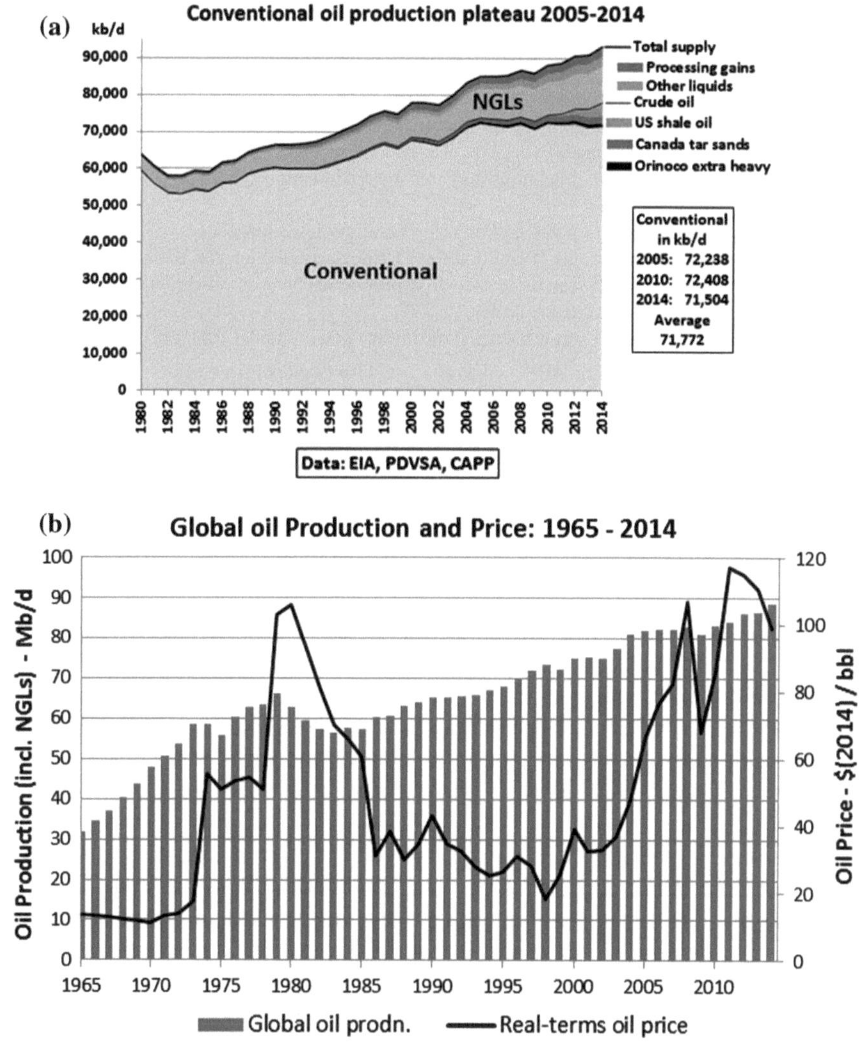

Overall, on these forecasts and data, and on the information presented earlier, we make the following judgements on the likely future global production of different categories of oil. Note that such judgements recognise the large uncertainty in the underlying data, particularly on Middle East and FSU reserves for conventional oil and on the energy-return data for the non-conventionals.

(a) World production of 'Regular conventional' oil, as defined by Campbell

It seems fairly clear that the production peak of this oil is already past, most probably in 2005.

◀ **Fig. 4.9** **a** Global 'All-liquids' Production by Category of Liquids, 1980–2014.

Notes

- The data are:
  - For most categories of liquids, and total, from http://www.eia.gov/cfapps/ipdbproject/IEDIndex3.cfm.
  - For tar sands oil (in Canada): http://www.capp.ca/publications-and-statistics/publications/258990.
  - For Orinoco oil: http://www.menpet.gob.ve/secciones.php?option=view&idS=21.
- Note that here Mushalik notes "I could not get Orinoco production for 2013 and 2014. I just took the 2012 production [for these dates]. If production were a bit different, it would not show up in the graph due to the scale".
- Production of GTLs, CTLs and biofuels are in the 'other liquids' category.

Source Chart by M. Mushalik of ASPO Australia, see http://crudeoilpeak.info.

b Global 'All-oil' Production (*vertical bars, left scale*), and Annual-average Oil price, Real-terms ($2014/bbl, *line*, *right scale*); 1965–2014.

Source BP *Statistical Review 2015*; based on an original plot by E. Mearns

(b) World production of 'all-conventional' oil (i.e. of total crude oil less light-tight oil, the extra-heavies of tar sands and Orinoco oil, and also excluding NGLs)

The peak of 'all-conventional' oil is probably about now at current oil prices. This peak is driven primarily by 'below-ground' recoverable resource limits (i.e., not enough oil in fields having been discovered, combined with the physics of field size distribution and field decline). But 'above-ground' limits also play a part, including resource nationalism issues (access, and willingness to produce) in some producer countries, and supply difficulties in specific countries including Libya, Iraq and Venezuela.

The above-ground constraints may change, but if in a direction to increase production, experience and geopolitics suggest that other constraints are likely to replace them. In particular, as the world wakes up to the realities of peak oil, it is probable that more producer countries will be less willing to export, with early indications of this already apparent.

Thus we can say with reasonable confidence that the world peak of conventional oil production at current prices is about now; it is primarily 'resource-limited'; and it was the primary cause of the rise in the oil price since 2004.

If the oil price stays high undoubtedly more conventional oil will be produced, some by simple in-fill drilling, some by extra discovery, but mainly by increased application of EOR techniques. But the extra quantity of conventional oil will probably not be so very large, based on experience from past-peak regions such as the UK, where higher prices have impacted production only little. With the recent period of high oil price damaging economies, it is likely that the primary effect of a sustained higher price will not be significantly greater production of conventional oil but a reduction in demand, particularly in the developed countries.

(c) World production of 'all-oil' (i.e. 'all-liquids' less GTLs, CTLs and biofuels)

The forecast for 'all-oil' is less certain, as with sufficient investment and project approval by governments it is clear that the total of light-tight oil from fracking (though there is not so very much of this globally), tar sands and other heavies, plus a return to interest in kerogen oil, could see significant increases in production. However, whether this can be enough to offset the loss of conventional oil production, expected to be 1.5–2.0 Mb/d each year is not clear. If this annual increase in the production of the non-conventional oils does not come forward, the global peak of 'all-oil' production would be expected around ~2020 or so.

(d) World peak in the production of 'all-liquids'

Finally what about the 'all-liquids' peak? Certainly, here the total resource base, of biofuels, CTLs (though not GTLs, unless UCG is included) and kerogen oil (included above in all-oils) is large and imposes no significant near-term supply constraint. But the intrinsic costs due to relatively low EROI ratios (for GTLs, CTLs, kerogen oil, and biofuels), competition with agriculture and biodiversity (for biofuels), probable CO_2 taxes and other CO_2 limits that may be mandated (for all fossil fuels) and the impact of a high oil price on global economies all make rapid increases in production of these fuels, adequate to offset conventional oil's decline certainly questionable.

Therefore—though forecasters need to expand their models to include the above factors, especially for energy return, if the world's oil future is to be properly understood—the conclusion is that the world almost certainly now faces an increasingly constrained, and quite possibly a peak, in total 'all-liquids' supply.

Chapter 5
Why Has Peak Oil Been so Poorly Understood?

Finally, in this introduction to peak oil, we need to look at the question of why the concept of peak oil has been so poorly understood.

This topic is important, because for many years the 'mainstream' forecasters (such as the IEA, US EIA, OPEC, most oil majors, and most consultancies) gave no warning of the recent steep rise in oil price. Moreover, most of these authorities specifically declared that concerns about the global conventional oil peak to be unjustified (see below, and Annex 5).

This failure by the mainstream modellers meant that the long history of 'technically-based' warnings of global oil supply limitations being expected around the year 2000 or shortly thereafter was not incorporated into societal, government or business thinking. As noted earlier, the high oil price that resulted from proximity to the global conventional oil peak returned the price to that of the 1978 oil shock in real terms (see Fig. 4.9(b)) and helped trigger the 2008 global recession, mirroring the 1970s price rises and the recession of the 1980s.

Moreover, the current on-average high oil price is still impacting economic growth. Such growth is almost universally seen as desirable, both to defray past debt (some from the recession itself via reduced tax receipts, increased welfare costs and bail-out and quantitative easing costs), and to raise future standards of living. Had the warnings of the risk of a near-term high oil price been properly recognised, a wide range of decisions by society, business and government could have been better taken.

The main reasons for the errors in the 'mainstream' oil forecasts over the years have been as follows:

(i) Use of proved (1P) reserves data. As already explained, these data are very misleading, and particularly so in terms of apparent 'reserves replacement'. It is hard to overstate the significance of analysts being misled by proved reserves in the general failure to understand peak oil.

(ii) Reliance on generally-available current oil data to forecast production; specifically the production trend, size of reserves, recent discoveries and technology gain. Use of these data in combination would seem reasonable, but from these data alone a proximate peak cannot be forecast, and indeed is counter-intuitive.

R.W. Bentley, *Introduction to Peak Oil*, Lecture Notes in Energy 34,
DOI 10.1007/978-3-319-26372-4_5

(iii) Not appreciating the significance of the 'mid-point' peak of conventional oil when assessing future supply.
(iv) Use of URR estimates substantially larger than indicated by extrapolation of 2P discovery.
(v) Adoption of an 'economic' view of resource availability.
(vi) Confusion as to which category of oil is being forecast.
(vii) Lack of analytical input.

These reasons are discussed in turn.

5.1 Use of Proved Oil Reserves Data

Chapter 1 and Annex 2 explained why the 1P and 2P oil reserves data are so very different. But as mentioned, more significant than these differences has been the fact that nearly all analysts and oil commentators have pointed to the *steady rise* in global proved (1P) reserves as proof that no oil supply constraints could be in sight.

For example, Odell (1997) used the large increases in global proved reserves as reported by DeGolyer and MacNaughton at that time to state that the world was "*running into oil*", not out of it. (This increase was almost entirely the OPEC ~300 Gb 'quota wars' increases of the late 1980s, the significance of which was apparently not known to Odell.) Adelman (also in 1997) protested similarly that "*World-wide discovery rates are said to have dwindled for 35 years. Yet production and proved reserves ... are at record levels. We hear only famine, and we see only plenty*". Watkins (2006) repeated the same mistake of misunderstanding the apparent increases in proved reserves in his article: *Oil scarcity: What have the past three decades revealed?* and likewise Radetzki (2010) in *Peak oil and other threatening peaks—Chimeras without substance.*

Even oil companies are not immune from this error of misunderstanding apparent 'reserves replacement'. Peter Davies, former Chief Economist at BP told a UK House of Lords Committee (2002): "*... we have continually replaced both oil and gas reserves over the past 30–40 years*". And the company's recent Group Chief Economist, Christof Rühl, wrote "*... all the global "end of oil" predictions so far have been wrong. Proved reserves, for example, more than doubled since 1980—despite the world consuming more oil over the last 34 years than it had in proved reserves in 1980*". (For both these quotes see the 'BP' section in Annex 5.) BP is not alone in promulgating this view; a few years back the CEO of Chevron dismissed any fear of global oil supply difficulties on the basis that "*world proved reserves continue to increase*" (see also Annex 5).

A further recent and egregious example of this error was that by Steven Chu (then US Energy Secretary) and Arun Majumdar in a 'Perspective' article in *Nature* (2012) where they state in their Fig. 1 that "*Although the [global] consumption of oil has increased 31 % between 1980 and 2008 the known reserves have increased comparably owing to improvements in exploration and extraction technologies*", and giving the source for this as BP. For a high-profile article to be so misinformed on this key statistic is very disappointing indeed.[12]

However, in partial mitigation, it should be pointed out that a number of well-known public data sources, such as the *Oil and Gas Journal*, also seem not to understand the issue, and each year simply highlight the increases in global proved reserves over the previous year. For example, the journal's 2014 *Worldwide Oil Field Production* survey states "*Worldwide oil and gas reserves estimates have increased slightly from a year ago, according to OGJ's latest survey of proved reserves*". While this statement is technically correct, all published proved reserves data (such as in the BP *Statistical Review*) should be accompanied by a strong caveat warning readers of the many problems with these data, and in particular against drawing conclusions from their apparent growth over time.[13]

Thus overall, while mistakes about global 'reserves replacement' are perhaps understandable if only the 1P data are known, such views stand in stark contrast with the fact that the industry backdated 2P global reserves of conventional oil have been in decline for nearly 35 years as Figs. 2.21–2.23 show. With the 2P data being fairly hard to access, and even the *need to access* such data not appreciated by most analysts, it is little surprise that the technical forecasters using these latter data have had difficulty in communicating their warnings about the proximity of the global conventional oil peak.

5.2 Reliance on Generally-Available Current Oil Data to Forecast Production

Next, we turn from the reserves data to misleading conclusions that many analysts have drawn from a wider set of oil data current at any given date. It has been common to take a view on future oil production in the light of four current parameters: the production trend, size of reserves, current discoveries and gains from improvement in knowledge and technology. Unfortunately—and perhaps surprisingly—on this set of data a proximate peak in oil production cannot be forecast, and indeed is counter-intuitive.

This is illustrated in Fig. 5.1, which is based on the simple model given earlier in Fig. 2.4. As that had shown, if the first field in the model starts production in year 1 then the region as a whole reaches peak in year 12.

In Fig. 5.1 the question then becomes *What do most analysts see at year 10?*

- First, they see that production has been rising rapidly in the past, and is still trending upward. Though growth has slackened somewhat, given the variability of real-world production, no analyst on these data alone would predict an imminent *reversal* of the upward production trend.
- And what about reserves? Since the model assumes that fields are discovered 5 years before getting into production, the reserves at year 10 are as shown in Fig. 5.1 (right). Here field 1 still has nearly half its original reserves; fields 2–9

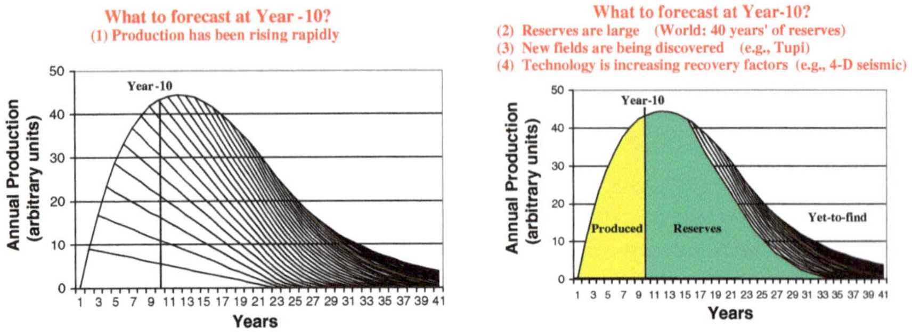

Fig. 5.1 The problem with using current oil data to forecast production: On these data the peak is *counter-intuitive*.

Left chart The field profiles, and hence regional profile, from the simple model of Fig. 2.4, with the situation at 'year 10' indicated by the *vertical line*. At this date production has risen rapidly in the past and is still rising fairly fast.

Right chart Components of URR at year 10:

- Solid area to the left of the year 10 line: The oil produced by that date: only a small proportion of the URR.
- Solid area to the right of this line: The large amount of oil in reserves, comprising that remaining in fields under production in year 10, plus that in the fields already discovered, i.e. those that will come on-stream in years 11–15. Though the peak is only 2 years away, reserves are enough for about 12 more years of production at the current production rate.
- Fields to the right of the plot: The numerous fields that remain to be discovered as of year 10.
- Not shown: Fields in production can increase in volume due to gains from knowledge and technology ('reserves growth')

have considerably higher fractions; field 10 is only just coming into production; and the reserves data must also include those in fields 11–15, which have been discovered but are not yet in production.

- Thus at year 10 there are large quantities of reserves remaining. Moreover, these reserves are mostly low cost. This is because most are in fields already in production, where the incremental cost of production—as long as kept within existing production profiles—is low, while reserves in the fields not yet in production are in a region where the geology is understood, and much of the infrastructure already in place.
- The data on the *discovery* trend at year 10 are also encouraging. While the rate of discovery is slower than at the beginning, discovery is still continuing with many new fields—albeit smaller in size—being discovered (Fig. 5.1, right, indicates the many fields in the model not discovered by year 10).
- And finally, technology moves on apace, with recovery factors in fields in production almost certainly increasing.

Thus this 'generally-available' set of current oil data typically indicates:

- production: ever-upward;
- current reserves: many years' worth;
- current discoveries: continuing, including, usually, some large late finds;
- knowledge and technology: increasing recovery factors.

And when this information is combined with a reasonable estimate of the *total volume* of recoverable oil expected to be available in the region (reserves, plus yet-to-find, plus technology gain), it would seem quite reasonable to forecast 'no peak in sight'. On these data the date of peak production of conventional oil in a region is not indicated at all; and indeed at the model's 'year-10' the proximity of such a peak would seem completely counter-intuitive to most observers.

We can see this more explicitly by looking at the real-world data for end-2004 (i.e., just before the price rise set in). At that date:

- Total global oil production was still rising strongly, at nearly 5 % p.a.
- Global conventional oil reserves stood at about 1200 Gb (whether measured as 1P or 2P), giving an R/P ratio of an apparently secure 45 years' worth of production.
- The global ultimate recovery (taking, e.g. the USGS year-2000 URR figure of 3345 Gb incl. NGLs and allowance for reserves growth) indicated that the likely remaining recoverable resource of conventional oil was considerably larger than remaining reserves, being over 2000 Gb, i.e. about 65 years' of production.
- Many large new oil discoveries were being reported, including in the Gulf of Mexico, offshore Brazil, offshore Africa, the Caspian region and elsewhere.
- Significant reserves growth was apparent in the industry backdated 2P discovery data, indicating continuing gains from increasing knowledge and technology.

On the basis of these data, most 'mainstream' forecasters in 2004 (i.e. at the equivalent of 'year 10' in Fig. 5.1) predicted that global conventional oil production would increase steadily in a 'business-as-usual' manner throughout the whole of their forecast periods (typically out to 2030). That is, forecasters cannot see a peak of conventional oil coming unless they (a) use the 2P data and (b) take into account the mid-point peak of conventional oil.

This latter point is discussed next.

5.3 Ignorance of 'Mid-Point' Peak of Conventional Oil When Assessing Future Supply

The significance of the 'mid-point' peak for conventional oil has been discussed above, but here three particularly influential examples are presented of where this concept was overlooked.

(Note that as explained earlier, 'mid-point' does not mean that production in a region peaks at *exactly the mid-point* of some particular estimate of the region's URR, but refers to the concept that production of conventional oil in most regions reaches a resource-limited peak when *roughly half or less* of the recoverable amount in the region has been produced.)

The first example is again of statements by BP's former Chief Economist, Peter Davies. On many occasions when questioned about peak oil he responded by citing the world's 40 years of proved reserves as proof that peak must be far away (e.g. in the House of Lords 2001 committee presentation mentioned above; see Annex 5).

It was always easy to show that this argument is misleading. At the date most of these statements were made, the world had consumed about 1000 Gb of oil; had a total original endowment (URR) of conventional oil estimated variously between about 2000 to 3000 Gb (excluding NGLs); and was consuming about 30 Gb/year. Using the lower URR estimate and a 'mid-point' calculation, the world was thus roughly at the conventional oil peak, and the entire secure 40 years of the proved reserves would be produced only at a declining rate, post-peak. With the higher URR, peak would be in about 17 years, and hence more than half the 40 years of proved reserves would be produced post-peak. It is not clear whether Peter Davies ever understood this concept, but his regular dismissal of the "peak oil theory" carried a great deal of weight in many circles, see Annex 5 (also for a discussion of 'peak oil' within BP).

A second example of apparent failure to understand 'mid-point' was even more influential. For many years almost all the 'mainstream' forecasts (including those from the IEA, US EIA, OPEC and the WETO study for the European Commission) modelled future oil *demand* in detail. But on supply they noted simply that anticipated summed demand over their forecast horizon was much less than estimates of remaining oil (which was correct, see Fig. 2.24). But this missed the concept of 'mid-point' for conventional oil, and as a result these organisations over many years carried out virtually no detailed supply modelling at all, and hence forecast no supply constraints (and hence also low oil prices) out to their forecast end dates. Figure 5.2 presents the evolution of IEA forecasts from 1998 to 2012, and shows the slow increase of realism in these forecasts as time progressed. (Importantly, the IEA *had* understood the conventional oil peak in the text of its 1998 *World Energy Outlook*, but dismissed the concept again until about 2008, see Annex 5.)[14]

A third example of apparently not comprehending the significance of mid-point are the studies which present estimates for the *total* global oil available (usually by category) such as Fig. 2.24, but which do not warn readers that the data must be viewed with the additional perspective of mid-point peak. Most of these studies, except for this key oversight, provide otherwise valuable data. Examples are the recent GEA/IIASA Global Energy Assessment (2012), papers by R. Aguilera, and the IEA's *Resources to Reserves*, 2008 edition.

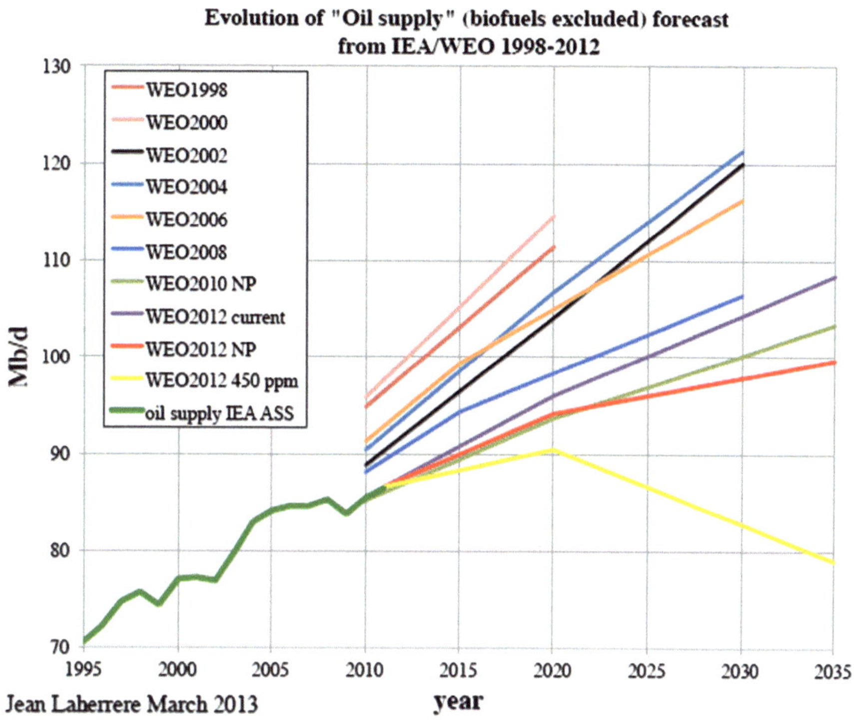

Fig. 5.2 Evolution of IEA *World Energy Outlook* forecasts for global oil production, published from 1998 to 2012, versus actual production.

Notes

- The text of the 1998 forecast (but not the Executive Summary) did point to an expected peak of conventional oil production, see Annex 5.
- In these forecasts field decline has probably only been explicitly modelled from 2008.

Source for the chart J. Laherrère

5.4 Use of URR Estimates that are Substantially Larger than Extrapolation of Discovery

Yet, another reason for the poor understanding of peak oil has been the over-reliance on 'assessed' estimates for the global URR (such as from the USGS, but also other organisations), and not giving sufficient attention to URRs indicated by extrapolation of the backdated 2P discovery trend. Since this mostly applies to analysis within 'mainstream' oil forecasting organisations it is discussed in Sect. A4.7 within Annex 4 on aspects of oil forecasting.

5.5 Approaching the Problem from an 'Economic' Viewpoint

A fifth reason that peak oil has been poorly understood has been the holding by 'non-peakists' of a traditional 'economic' viewpoint of resource availability.

(a) Oil reserves should be seen as 'just inventory'

Implicit in standard supply/demand curves is the assumption that at a higher price more supply is forthcoming. In oil's case, this process has been expressed as turning 'resources into reserves'; and since the global resources of all classes of oil and of 'near oil' (such as kerogen) are indeed very large (Fig. 2.24), the view has some merit.

But the counter-examples are the many regions past peak. For example, in the UK oil production peaked in 1999 when the oil price briefly fell to a spot low of ~ $10/bbl. Subsequently, the oil price rose tenfold to ~$100/bbl, but production fell. This resulted in a somewhat curious supply/demand curve, where both supply and demand fall as the price increases! Of course economics is familiar with products where as supply declines from one source another steps in, and oil is indeed fungible. But nevertheless the lesson from regions like the UK is key: conventional oil declines, and more is not available even at much higher price.

As already mentioned, the 'economic' view for oil supply came to prevail in part because of oil's history: in the early 1970s to early 1980s global proved reserves stood at about 30 years' of supply, and the widespread view was that the then-existing oil shortages would get ever more severe, with oil possibly 'running out' altogether at the end of the 30 years or so. (Why this mainstream view did not understand the need to add in the then-known large amounts of conventional oil in probable reserves, plus the yet-to-find, plus the likely gains from improvements in recovery factor, and also the known oil available from tar sand and kerogen deposits —as the 'technically-based' forecasters at the time knew—is not at all clear.)

When the oil price collapsed from the mid-1980s on, the consensus view switched dramatically: from expecting oil exhaustion to seeing the 'economic' view of resource availability as being correct. Moreover, as also explained, the data to support this view seemed to be at hand, as the (then relatively small) amounts of oil in *proved* reserves were apparently being continually replaced. In the US, for example, proved reserves have held steady at an R/P ratio of about 10 years of supply for now over a century, while in the UK proved reserves have stayed at an R/P ratio of about 5 years of supply for two decades.

Thus the 'Adelman' view came to prevail, where the quantity of oil available was "unknown and unknowable"; and that any forecast based on an assumed fixed quantity of recoverable resource was seen as methodologically grossly in error. (This latter view, in turn, was poorly received—not surprisingly—by exploration geologists who had spent their professional lives quantifying with some precision just how much oil was out there, region by region.)

However, some sympathy must be accorded this aspect of economic thinking: why were 2P reserves not treated as normal inventory, to be replaced as needed? In other words, why had the oil companies gone to the expense and trouble of exploring for more oil when there were already very large quantities of 2P reserves 'in the bank'? As Fig. 2.23 for example shows, these reserves reached over 140 years of supply once Ghawar was found (and with even higher '2P'R/P ratios being achieved in some earlier periods).

The answer is provided by Yergin's *The Prize*:

- Sometimes, large oil reserves were held by a company with a small market, and vice versa, so that a company in the latter category needed to look for more.
- Sometimes, it was simply the entrepreneurial spirit of 'independents' to look for more; originally, the relatively small independents in the US fighting Standard Oil, and then fighting the US majors that appeared (many spawned from Standard). But subsequently the 'independents' were the smaller (but still large) newer oil companies (ENI, for example) that were prepared to take on large exploration risks in order to break the market stranglehold of the by then very large incumbents.
- Also, as Colin Campbell notes, for commercial companies, exploration is fully tax deductible, and hence can be effectively nearly free to these companies.
- But almost certainly the main reason to keep looking for ever more oil, even when the global 2P reserves were already large, was that once the reserves were 'overseas' (outside of the US; or both within or outside Russia in the case of the main Russian suppliers), these reserves were not nearly as large as they seemed. Such 'overseas' reserves were always at the risk of higher rent, nationalisation or complete loss (from the oil company's point of view) from embargo, revolution or war. Yergin has good sections on how senior executives of the various oil majors judged these risks. Thus, counter to a pure 'economic' view, there was *always* a strong need for individual companies to find more oil.

Note, moreover, that this general economic view of peak oil in terms of 'resource availability' became intertwined with the grossly ill-informed controversy over the '*Limits to Growth*' study (for an explanation of this, see Bardi 2011); such that it became standard for critics of peak oil to say 'such a view just shows '*Limits to Growth*' ignorance'.

Looking back, it is a great shame that the two sides of the debate, essentially geologists and economists, could not find mechanisms to adequately explain their views to those in the opposing camp; see, e.g., the key IEA 1997 Paris meeting as reported by various attendees in Campbell (Ed.) (2011).

(b) Other economic arguments

In addition, other general economic arguments have also significantly impacted the 'peak oil' debate at various times. These have included:

- 'There is no need to worry about an oil peak, because once the market senses that any future risk of supply constraints might be genuine this will be reflected in the value of reserves'. Since examination of the data for many years showed that reserves had not increased significantly in value, any risk of near or medium term oil difficulties could not be real. (The counter-argument is simple: since few of the 'mainstream' forecasters understood the impact of conventional oil's mid-point, it is not surprising that risk of future oil scarcity was not being priced into reserves.)

- There was also little need to worry about an oil peak, since for any resource there is always a 'backstop' resource. (The flaw here was that the scientists could not—and still cannot—think of any currently large, easily available, good EROI, cheap alternative to oil for powering large and critical parts of the global economy.)
- Economists pointed out—quite correctly—that mineral supply is a race between resource depletion and a combination of inventiveness, rising energy use, and investment that allows ever lower grade sources to be extracted. This view then said that since for nearly all minerals their history over the last century or more had shown this race was being won, with the real terms cost of such minerals having fallen rather than risen, there was no reason that this principle was not also true for oil. (This is a plausible argument, but the reply is the detailed arguments for peak oil set out above. And on minerals generally, see Bardi 2014.)

Many other reasons, not solely economic, were also put forward to show that peak oil was obviously a 'myth' (a preferred term among most of the 'non-peak' authors). These reasons included the technological progress in oil recovery (Kern River and Weyburn being cited, as mentioned earlier); that there were still many places not fully explored (quite true, but *how prospective* in the view of knowledgeable exploration geologists?); and the very low number of exploration wells in Middle East countries compared to wildcats in the US (also true; but as already mentioned, the Middle East's large structures need few exploration wells once the seismic is shot, while the US' land ownership rules, generally more complex geology, number of basins, and wide range of field sizes, lead to the need for many).

5.6 Confusion as to Which Category of Oil was Being Forecast

Next, in this section we address the confusion that has sometimes occurred as to which category of oil is being forecast.

For example, Christof Rühl in his recent blog 'Peak Oil' mentioned above (Rühl 2014) wrote "*How has the original* [Hubbert] *Curve been holding up? ... We can be confidant now that Hubbert's prediction for US oil production will be out by an order of magnitude*".

Rühl's mistake here is to not recognise that Hubbert's prediction for US production was *explicitly* for only conventional oil production, ex-NGLs, and only in the Lower-48 states. So Alaska, deep offshore, very heavies, EOR, light-tight, NGLs, kerogen oil and biofuels were all explicitly not included (see Annex 5). On US conventional oil production, Hubbert's prediction is surprisingly accurate even 40 years later; even though his purpose was certainly *not* to forecast production 40 years on, but to alert Americans to the near-term dramatic transition in their oil supply, from production going up to production going down.

Rühl might reply 'Even so, US production *is* higher than Hubbert's forecast', and this is true. But this is to miss an even more important point. As already noted, the US peak in 1970 had the effect of allowing OPEC, who had long wished to see a higher price (and to this end had tried production cutbacks before) to finally control the price of oil; as a result the oil price rose in real terms to today's level, and a decade-long global recession resulted. This high oil price caught out the public, politicians and analysts alike, despite Hubbert's clear warning of a coming sea-change in the US contribution to global oil supply made 15 years before. In a parallel fashion, the high on-average oil price since about 2005 was almost completely unanticipated by all mainstream organisations, despite 50 years of clear technically-based warnings that global conventional oil production would peak, and hence more expensive oil be required, from around the year 2000.

Rühl's error, therefore, is not to understand where forecasts have been made for specific regions and types of oil, and thus should be judged only against these. And moreover, that the larger 'all-liquids' (and indeed, all-energy) picture will only emerge from a proper understanding of its components.

5.7 Lack of Analytical Input

To conclude this section on why peak oil has been poorly understood, it should be pointed out that one of the key factors has undoubtedly been the comparatively few analysts working quantitatively in this area, at least since the late 1980s.

For example, the University of Reading was long the only UK academic group looking quantitatively at the problem; as far as academic or consultancy research effort went, future global oil supply was essentially just not 'on the agenda'.[12]

Finally, despite a number of excellent books and papers on the topic, there have been relatively few 'beginner's guides'.

But in the end this is no excuse: the data and methods were well understood by a small number of technical specialists, and were widely published. There is thus more to be done to understand why such a fundamental and important lacuna in the understanding of otherwise competent analysts, across many organisations including government, industry and universities, could persist for so long.

Chapter 6
Caveats and Conclusions

In this chapter, we present some caveats to the information provided above, and also the main conclusions to be drawn.

6.1 Caveats

A book of this nature calls for many caveats; some are fairly straightforward.

(a) Quantities

Quite a number of the phrases used in book, and especially in Chap. 1 (such as "the quantity of currently available biofuels falls very far short of what is needed to replace oil"), are unquantified: for example, what does 'very far short' mean in this context?

Such statements, although they might well be correct, unless quantified risk being simply assertions. As David Mackay points out in his book *Sustainable Energy: Without the Hot Air* (Mackay 2009), quantitative data are needed to support almost any valid argument.

Here the units to use are exajoules (EJ), where the global annual traded energy is currently about 550 EJ/y, and hence oil makes up about 180 EJ/y of this. On a *net-energy* basis, oil contributes perhaps something around 170 EJ/y; and it is this quantity of annual net-energy that biofuels would need to match to fully replace oil. Quantitative data on the potential availability of nearly all current energy sources are given in the excellent *Global Energy Assessment* study (IIASA 2012); and see also Mackay's book, just mentioned, for data on the energy from renewables.

(b) The possibility of significant new sources of energy

In the chapters above we have painted quite a bleak picture on the world's energy future, at least in the near and medium term. However, it is always possible that one or more excellent new sources of energy will come out of the research laboratories that would change the picture entirely.

R.W. Bentley, *Introduction to Peak Oil*, Lecture Notes in Energy 34,
DOI 10.1007/978-3-319-26372-4_6

(c) Discovery of conventional oil

Similarly, discovery of oil has always contained surprises. While it would seem hard to turn around the half-century of decline of discoveries of oil in new fields when set against the context of an ever-increasing knowledge of how such oil is formed and located, if such a turnaround were to occur, again the picture would change.

(d) Recovery factor

The picture also changes if splendid achievements were to take place in the percentage of oil that can be recovered from conventional oil fields. By volume average, this percentage is currently only around 40 %, that is, current technology, and the current price, typically leave some 60 % of the oil in fields unrecovered. This percentage will surely rise, though many think by not so very much. If a step-change were to occur in global-average recovery factor, the global oil supply picture would look far more optimistic.

(e) Reduction in oil consumption

For reasons of CO_2 emissions, or high price, or advancement in technologies for the efficient use of oil, the global use of oil might decline faster than the rate of decline in supply that is suggested here. This would change the overall global oil supply/demand balance for the better; and is briefly discussed in Annex 7 on 'peak demand'.

(f) Are the oil supply changes expected in the 'Second half of the Oil Age' so very different from earlier changes in oil price and supply that have occurred in the past?

Followers of the oil situation know that the supply of oil has always been in a state of flux; it is sensible to ask if the coming change is simply not one of a similar magnitude.

In the past, supply has migrated from shallow wells onshore in mainly Russia and the US to encompass oil from an ever-widening number of supplier countries; and to oil initially from ever-deeper wells, then from offshore fields and to heavy oils that need thermal stimulation to produce; then to oil from the Arctic, and from 'pré-sal' fields off Brazil; and, since the 1960s, to tar sands oils and the increasing range of non-conventional oils including, most recently, shale oil.

So this is an important question to ask. As this book has tried to show, the defining feature of 'The Second half of the Oil Age' is the decline of oil from *oil fields*, and hence the need to increasingly supply the world's oil from the generally geographically extensive, more intrinsically difficult to extract, sources of oil. This has already given a step-change upwards in the price of oil, but only time will tell for certain if this turns out also to give the step-change downwards in the supply of oil that this book suggests will be the case.

(g) And what about the short term?

In the short term, since oil is now partly back to being a simple traded commodity (i.e. where typically a slight over-supply leads to a significant price drop, and slight under-supply to a significant price rise), and since some exporters still have spare capacity that they can put on the market if they feel conditions dictate, then large price volatility is to be expected; as indeed has been the case at the time of writing (Sept. 2015) since about 2005. But as the potential for global over-supply of conventional oil diminishes, the price of oil would be expected to come to rest—on average—fairly close to the cost of the marginal barrels.

6.2 Conclusions

With the above caveats in mind, this book concludes that:

1. The mechanisms that produce a resource-limited peak in oil production are different for a field, a region and globally; and for conventional oil alone versus for all-oils and all-liquids; see Chap. 2.
2. Proved oil reserves ('1P') data cannot be used in general to predict future oil production; instead industry proved-plus-probable ('2P') data must be used. This is because:

   - '1P' data are generally *underestimates* of the quantity of oil likely to be extracted, being normally less than the more likely proved-plus-probable ('2P') reserves. This was especially true in the past, where SEC rules meant that large fields only reported oil in communication with existing wells, and hence proved reserves increased as fields were drilled up over time. In addition, proved reserves for *regions* are also smaller than 2P values because of incorrect statistical aggregation. It was the misleading impression of 'reserves replacement', often driven simply by growth of 1P values towards 2P, which led many analysts to conclude that 'reserves are just inventory', reserves replacement was the norm, and that the total amount of oil available was "unknown and unknowable". With the passage of time, however, the total of proved reserves in many regions has now grown fairly close to the corresponding 2P values.
   - In certain Middle East OPEC countries proved reserves are probably *over-estimated*, being significantly larger than the corresponding 2P values held in industry databases. This arose largely from the 'quota wars' manoeuvring for production volumes in the late-1980s; and may represent a total overstatement of these reserves by ~300 Gb.
   - Proved reserves data for many counties are often *not estimated* on an annual basis, and hence remain unchanged, sometimes for many years in a row. This also limits the value of conclusions drawn from changes in these data.

3. An R/P ratio gives no indication of future rate of oil supply.

A common forecasting error lies in using a reserves-to-production (R/P) ratio to indicate a region's security of oil supply. Such a ratio, whether based on 1P or 2P data, contains no information whatsoever about the region's future *rate* of production, because the resource-limited production peak may be already past or close. Today, if counting offshore separately from onshore, there are perhaps a hundred major geographical regions that are past their resource-limited peak of production of conventional oil but where their R/P ratios give no indication of this. Likewise, for the world as a whole, the R/P ratio remains reassuringly large, at over 50 years of production, despite the fact that the resource-limited peak of global conventional oil production (ex-NGLs) is close or already past, depending on the precise definition of 'conventional'.

4. The 'resource-limited' peak in the production of conventional oil (essentially, oil in fields) in a region occurs because not enough of this type of oil has been found, coupled with the physics and economics of production.

The primary driver for the peak of conventional oil production in a region is that not enough oil has been found in fields. The peak in production occurs sometime after the peak of discovery, and occurs at the point where the decline in production from the—generally larger—early fields is no longer compensated for by increasing production from the—generally smaller—later fields.

The main physical characteristics that drive this peak are the field size distribution in the region (where smaller fields not only yield less oil, but also tend to be harder to find), and the decline in production from individual fields, where the latter is typically due to some combination of pressure loss, reduction in oil column and increasing bypass of the drive fluid.

The economics of production of conventional says that once a region is in its 'resource-limited' decline, it would require a very large amount of investment to increase production from the large existing fields already in decline, and to find and bring on-stream a large number of smaller later fields. At oil prices that have been typical over recent years, about $100/bbl, a rule of thumb might be that trying to reverse production decline of *conventional* oil in a region past its peak is not in general possible at these prices; production from the US, UK or Norway providing examples.

An early warning of this conventional oil 'resource-limited' production peak in a region is signalled by the tapering-off in the rate of conventional oil discovery in the region, which, as mentioned above, must be evaluated from the industry 2P data.

In regions where above-ground constraints are not dominant, both theoretical modelling and extensive empirical evidence show that the resource-limited conventional oil peak occurs typically at, or somewhat before, the 'mid-point' of the region's apparent 'ultimate' recovery ('URR'), provided the latter is estimated from extrapolated discovery, and where also petroleum geology should be consulted to indicate if further significant plays are likely, and reservoir engineering used to indicate the scope for extra oil from the application of EOR techniques.

If a higher 'assessed' estimate for the URR for a region is used, particularly if this includes significant allowance for 2P reserves growth, then the peak of conventional oil production in the region typically occurs well before the 'mid-point' of this type of URR estimate.

5. The expected date of the global *conventional oil* peak in production is about now.

The data and models presented in this paper indicate that the world is close to its peak of conventional oil production at current oil prices, with the date of this peak being between perhaps 2005 and 2020, depending in part on how conventional oil is defined.

The peak results from a combination of below-ground resource limits and above-ground 'political' factors, but below-ground limits dominate. Data for the production costs of the marginal barrels required to meet global demand due to the slowing production of conventional oil show that proximity to the conventional oil peak has been the primary cause for the oil price to reach around $100/bbl.

6. Unknowns in forecasting the conventional oil peak.
   The main technical ('below-ground') unknowns in forecasting the conventional oil peak are:

   - size of Middle East and FSU 2P reserves;
   - rate that currently fallow fields will come on-stream for technical and political reasons;
   - potential gains from reserves growth in existing fields resulting from an extended period of ~$100/bbl oil; and how the timing of these gains will affect field and regional production profiles.

   The main 'above-ground' unknowns are:

   - risks of war, terrorism or similar;
   - impact of a high oil price on oil demand, and on wider global economic activity;
   - risk that oil price fluctuations will delay investment;
   - willingness of large exporters to produce as fast as high levels of investment would permit;
   - demand factors not related to price, such as imposition of CO_2 limits.

7. The impact of oil price and technology to increase the future production of conventional oil should not be overestimated; and the past *technically-based* 'fixed-resource' forecasts over the last approximately 40 years that have assumed a global URR for conventional oil production have proved generally correct.

A higher oil price and advances in technology can certainly lead to the discovery of more oil, access to oil in previously challenging places and an increase in recovery from existing fields. But it is easy to overestimate these impacts; and in terms of predicting the conventional oil peak a 'fixed-resource' model has proved to be a reasonable approximation.

Forecasts from the early 1970s to mid-1980s used estimates for the global URR for conventional oil, excluding NGLs, that mostly lay the region of 1800–2500 Gb, and combined these with a 'mid-point' approach to predict that the global oil peak or plateau of conventional oil production would occur around the year 2000.

Today, the corresponding conventional oil (less NGLs) URR estimate, if based on extrapolated 2P discovery, is roughly similar, lying in the range of 2300–2500 Gb; (and perhaps 300 Gb lower if lower estimates of likely 2P reserves are used, giving a range of 2000–2200 Gb). As a result, the expected date of the global conventional oil peak has shifted little from the forecasts of the mid-1970s to mid-1980s, particularly when the delay to the peak due to the fall in global demand due to the high prices of the 1970s and early 1980s is included.

It may seem surprising that there has been little change in the estimated global URR for conventional oil between the early 1970s and today, given the many new large oil discoveries in places like Gulf of Mexico, pre-salt Brazil, offshore West Africa, Kazakhstan and so on. The explanation lies in the fact, insufficiently recognised, that the 2P peak of *discovery* of conventional oil in new fields occurred in the mid-1960s, some 50 years ago, with new-field discovery declining ever since. It was thus relatively easy back in the early 1970s to extrapolate the past discovery trend to estimate expected future discoveries (oil that in fact has now largely been discovered), and hence to arrive at an estimated global URR for conventional oil in fields that has changed little since.

Thus as a result, and by contrast to the widely held view that 'all oil forecasts have been wrong', forecasts that the peak in global production of *conventional* oil (ex-NGLs) would occur roughly around the year 2000 have been correct for over 40 years.

Note that in predicting this peak these 'technical forecasts' combined a well-grounded method ('peak at mid-point') with good data (2P data), or later combined detailed by-field modelling also with good data. Both these approaches properly account for the two critical factors necessary to forecast production of conventional oil: the nature of the field size distribution, and the production decline in individual fields. By contrast, many of those that questioned peak oil did so because they combined a poor model ('reserves are just inventory', which ignores field size distribution and field decline) with poor data (1P reserves).

8. Forecasting production of *non-conventional* oil.

The dates of global peak in the production of 'all-oil' and 'all-liquids' are not so clear, and probably require more detailed modelling than is currently the case. This is because for non-conventional oil and all-liquids production, consideration is needed not only of the recoverable resource volumes, investment levels required and extraction costs, but also of technological readiness, EROI ratios, net-energy limits to deployment rate, public-acceptability and CO_2 limits.

Once fully past peak, the global production of conventional oil is expected to decline by roughly 2 Mb/d each year. As a result, and given the factors just mentioned, it seems likely that the global peaks in the production of 'all-oil' and 'all-liquids' may also occur reasonably soon. If this is the case, humankind will be increasingly constrained to curb its demand for oil due to these limits, as well as to avoid predicted severe climate change.

9. Modelling the energy future.

Finally, in these conclusions, we need to ask: What of the world's broader energy future?

There are a number of large energy resources on the planet, each very much larger than our current annual rate of commercial energy use. These include solar energy, deep geothermal, coal in thin and deep seams, oil from kerogen, uranium fission if breeder, possibly thorium in a fissile process, fusion and possibly methane hydrates.

Moreover, good design and planning can significantly decrease the demand for energy without loss of utility. For example, good insulation can keep a building warm for a very low energy spend; transport does not require much energy intrinsically; and the energy used in the manufacture of products can be significantly reduced by better design and recycling.

But to properly understand mankind's overall energy future, a global model needs to be explicit on the dynamics of energy flows (including peak conventional oil, peak conventional gas and possibly peak cheap coal; and also on EROI ratios, and net-energy rate limits); on monetary flows and investment sums available; and on aspects of demand management, including those to restrict CO_2 emissions.

Probably, no such comprehensive modelling is currently being done. It is needed if the world's energy future is to be understood.

Annex 1
Definitions

Some of the terms concerning peak oil are not yet fully settled. Here we use the following definitions, but for amplification see also the UKERC *Global Oil Depletion* Technical Report 1, (Sorrell and Speirs 2009), and the definitions given by Jean Laherrère on the ASPO France website: http://aspofrance.viabloga.com.

Conventional oil: Defined here as light and medium oil that has migrated from its source rock to a reservoir rock; usually having an oil–water contact; and where extraction is by primary recovery (own pressure, or mechanical pumping) or secondary recovery (natural gas or water drive). For reasons of the data already in some industry datasets, this definition can also include condensate, heavy oil produced by fairly standard thermal techniques such as 'huff and puff' and some oil currently (or expected soon to be) produced by enhanced oil recovery. Published data on conventional oil *production* often include natural gas liquids (NGLs), but in this paper the latter, where possible, are treated separately, as they come from gas wells having their own exploration history and production pattern.

> Note 1: Campbell uses the term 'Regular Conventional' oil to exclude all very heavy oil (<17.5 API), oil from >500 m water depth, Alaskan and other polar oil and NGLs. For conventional gas, he excludes the non-conventional gases (coalbed methane, shale gas, gas from underground coal combustion, and hydrates, etc.). The reason for these choices is that these excluded categories require different extraction and production profiles for modelling. Campbell notes that Canada has the cut-off for heavy oil at 25 API, and Venezuela at 22 API, but that he adopts a lower cut-off for heavy oil at 17.5 API so that "*all fields that can be produced in more or less normal ways may be included as 'Regular Conventional'*".

> Note 2: In general, some of the gas from gas fields condenses at the surface, called 'condensate', and some liquids can be produced by processing, 'natural gas plant liquids' (NGPLs). Thus, as a simplification: NGLs = NGPLs + condensate; and where—very roughly—NGLs currently contribute ~12 Mb/d to global liquids production, of which, the NGPLs perhaps ~9 Mb/d. But this is a confused topic, as the definition of, and inclusion within datasets, of these liquids are far from consistent (in part because some liquids can be identified and

R.W. Bentley, *Introduction to Peak Oil*, Lecture Notes in Energy 34,
DOI 10.1007/978-3-319-26372-4

sold, and others are blended into oil) and all 'gas liquids' data should be treated as approximate.

Enhanced oil recovery (EOR): Oil produced from an existing field by application of a tertiary recovery method such as N_2 or CO_2 injection, or use of other miscible liquids or chemicals. This oil is generally not seen as 'conventional' oil, as tertiary processes often—but by no means always—are used fairly late in a field's life (although this is now changing). EOR methods often incur significantly increased dollar and energy cost, and though they can be very effective in specific fields typically—at least to-date—yield relatively small extra volumes *in total* compared to the oil the fields would be expected to yield without their use, in the range of perhaps 5–15 % extra oil; see Fig. 2.24, and the section 'Reserves growth' in Annex 4.

Extra-heavy oil: In the main, refers to tar sands, other bitumen, and Orinoco oil, but includes very viscous oil also. (Note that Orinoco oil is in this category partly because of production constraints, as well as intrinsic oil characteristics.)

Light-tight oil ('shale-oil'): Oil extracted from rock that otherwise would be too impermeable for useful amounts of oil to flow to the (usually horizontal) borehole, by the use of high-pressure hydraulic fracturing combined with the use of 'proppants' that keep the fractures open despite the weight of overburden naturally trying to close these.

Oil from kerogen ('oil shale' oil): Oil produced oil shale and similar rock types that contain significant amounts of the oil precursor kerogen, and from which oil can be generated by retorting—either above ground or in situ.

'All oil': Conventional oil plus NGLs, EOR, extra-heavy oil, light-tight oil and oil from kerogen.

'All liquids': 'All-oil', plus gas-to-liquids, coal-to-liquids and biofuels.

Reserves: That quantity of oil or gas that has been discovered and is assessed as likely to be recovered under current or reasonably expected technical and economic conditions. This is the value usually held in industry databases as 'proved-plus-probable' ('2P') reserves.

As explained in Chap. 1, and in more detail in Annex 2, the public-domain *proved* ('1P') reserves data *do not* meet this definition; proved reserves data are very misleading, and cannot be used for oil forecasting unless additional assumptions are used (Bentley 2015a).

'Reserves' usually refer to the reserves *remaining* at a given date; while 'original reserves' refer to the quantity in a field or region before extraction started. Often data on the latter have been updated to reflect *today's* knowledge, so may not match the value published when production started.

Reserves growth: Refers to the change (usually, but not always, an increase) in the estimated ultimate recovery (URR, see below) that occurs over time as fields are developed and produced. See Annex 4 for a discussion of this issue.

Resource: All of a specified hydrocarbon in-place, whether discovered or not, and whether technically or economically recoverable or not. (Even this definition is not without ambiguity; as Miller asks: Does 'resource' include every molecule of the hydrocarbon in the rock, or just in sensible accumulations?)

The report *Undiscovered Petroleum Potential* by Laherrère, Perrodon and Demaison (Petroconsultants 1994) estimated that of all hydrocarbons generated, perhaps only 1 % will be classed as reserves; the remaining 99 % having been lost to the surface, or be unrecoverable in fields, or be still in the source rock, often trapped in fractures. For generated gas, some of this has been adsorbed into shale and coal. Here we define 'resource' as the entire hydrocarbon in a specified region; but use the term in a more restricted sense in 'resource-limited', as defined below.

Recoverable resource: That fraction of the resource that can be recovered under some stated or implicit level of technology and price. As mentioned earlier, for conventional oil (see the definition above) the global average quantity currently considered economically recoverable in existing fields is—very roughly—about 40 % of the in-place; while for conventional gas this ratio is—very roughly—about 80 %.

Ultimately recoverable resource ('URR'); also called estimated ultimate reserves ('EUR'): This is usually taken *not* as referring to some 'true ultimate' figure. (The latter, at least for a region, is probably impossible to know. Though limited by the original oil-in-place, which can in general be estimated, the ultimate amount recoverable depends on future price and technology, which are not known.) Instead, URR usually refers to the quantity of oil or gas judged likely to be extracted from the field or region by some distant date; the latter is sometimes not specified, but where specified can be 2070 or 2100, for example.

For a field its URR is the estimate of the total amount of oil the field will have generated at the date when production stops. As explained above, this estimate can change with time as changes occur in the knowledge about the field, in the technology employed to extract the oil or in the price secured for the oil.

For a region, allowance must also be made for the oil that is yet-to-find; and the equation becomes:

$$\begin{aligned}\text{URR} = {} & \text{Cumulative production to-date} + \text{Proved-plus-probable reserves} \\ & + \text{Expected reserves growth} + \text{Yet-to-find}.\end{aligned}$$

Resource-limited supply: As used in this book, the term applies to the production of *conventional* oil. It does not mean that supply is limited *solely* by the total quantity of the resource in a region, but that other resource characteristics, including current recovery factor and extraction techniques, and reasonably expected future technology and price, also act to limit the rate of supply.

For example, in this terminology, the UK peak of conventional oil production in 1999 is classed as resource-limited, as by 1999 not enough new oil in later fields had been discovered in the UK to offset the decline in the large fields discovered earlier. This production limit occurred despite the fact that the UK had then (and still has) *very large amounts* of flowable oil remaining in the ground, including in existing fields due to limits on recovery technology, in fields discovered but too small or otherwise difficult to be exploited economically at the current oil price, and in fields not then discovered. (And in terms of 'all-oil', the UK has in addition significant amounts of oil in rock too tight for migration, as oil shales and similar; and also in the form of kerogen.)

[At the risk of taking the argument in a slightly ridiculous direction, to make this idea clear suppose that oil from the UK North Sea was absolutely vital for some non-energy purpose. Then cofferdams could be constructed over oil fields, the reservoir rocks mined and pulverised, and every drop of oil extracted by chemical or thermal processing. In such a situation, and with enough resources available, UK oil production might exceed the rate in 1999. But if the oil is to be used as an energy source, and under any reasonable assumptions, then the UK's peak of *conventional* oil production in 1999 was indeed *resource-limited.*]

'Mid-point' peak: Applies to conventional oil, and refers to the fact that in a region where production has not been overly constrained by above-ground 'political' factors, production will usually peak at or before the point where roughly half the total recoverable resource (URR) of oil in the region has been produced. When referring to forecasting, the term applies to any technique where such a 'peak at *roughly* half way' is modelled, and hence includes use of a Hubbert curve, addition of multiple Hubbert curves, Campbell's approach, PFC's '60 %' rule, and also bottom-up by-field models (where the approximate 'mid-point' results from summing fields that themselves have, usually, early production peaks).

The approach is in contrast to forecasting approaches (or expectations, such from the use of R/P ratios, or 'cumulative availability curves') where no such explicit constraint is assumed on how fast the remaining recoverable resource of a region can be produced.

Units: Here the main unit of oil volume is US barrels (abbreviated as 'bbl' or 'b'). Production is usually in millions of barrels per day (Mb/d), while reserves and URR data are usually given in billions of barrels (Gb). No excuse is made for use of this distinctly non-SI unit, 'Gb', as it is so widely used in this field.

To understand peak oil, it is helpful to get a feeling for typical Gb data. In round terms, for example:

- Global oil production is ~30 Gb/yr.
- For conventional oil ex-NGLs, the UK's URR is ~30 Gb; the US Lower-48 URR is ~200 Gb, and the global URR is somewhere in the range ~2000–3000 Gb.
- Among the non-conventional oils, the global URR for 'light-tight' oil maybe about 250 Gb while that for oil from kerogen is ~1000 Gb.

Annex 2
Oil Reserves Data: 1P Versus 2P

This Annex expands on the information given in Chap. 1. As indicated in that chapter, a critical distinction to make in understanding peak oil is between *proved* ('1P') oil reserves data, and *proved-plus-probable* ('2P') reserves data (see Bentley et al. 2007); and where by far the largest part of the long-standing confusion over peak oil has resulted from the use of the former data.

A2.1 Proved Reserves ('1P') Data

Proved (often also called *proven*) oil reserves data ('1P' data) are widely available in the public domain. As mentioned in Chap. 1, the three main issues with these data are understatement, overstatement and non-statement. Here we amplify the information given earlier on these issues.

A2.1.1 Understatement

Historically, proved reserves have been *very much less* than " … those quantities that geological and engineering information indicate with reasonable certainty can be recovered in future under existing economic and operating conditions."; where the latter were far better indicated by the oil industry's proved-plus-probable (2P) reserves data.

(i) *US & Canada*

In the US and Canada in particular, the apparent size of individual fields if measured using 1P data has often grown significantly. Attanasi and Root (1994), for example, found that the aggregate size of US fields, as measured by cumulative production plus proved reserves at a given date, appeared to grow by sevenfold after 20 years of production, and by 13-fold after 90 years. Other approximations put US average field growth at about sixfold for onshore fields, and threefold for

R.W. Bentley, *Introduction to Peak Oil*, Lecture Notes in Energy 34,
DOI 10.1007/978-3-319-26372-4

offshore; while for Western Canadian fields, Odell found apparent growth in proved field volumes to be ninefold over field lifetimes.

There are many possible explanations for this very large increase over time in the apparent size of fields when measured by proved reserves, including, initially, ignorance of the true size of fields, and later for reasons of US or Canadian tax regimes; but the 'drilling-up' of large early fields is probably one of the major factors, as under SEC rules proved reserves can largely refer only to oil in communication with existing wells.

Attanasi and Root said there was insufficient US data to consider individual fields (and hence find detailed explanations), in part because US field data are proprietary. The authors raise a number of possible explanations for 'the enigma of field growth', and mention new pools, new reservoirs and the renaming of fields, but do not explicitly discuss drilling-up in terms of communication of wells to the oil. Note that the data showed, not surprisingly, even greater growth for old heavy oil Californian fields due to steam injection (such as Kern River, Fig. 2.2). In terms of the misleading issue of apparent reserves replacement when using proved data, the authors note that "… from 1978 through 1991, growth of old discoveries accounted for more than 90 % of additions to [US] proved reserves".

However, there is probably much still to be discovered as to why US and Canadian proved reserves data have been so conservative; and the subject also needs to be disentangled from true '2P' reserves growth, as discussed in Annex 4.

(ii) *The UK*

Though the US and Canada have probably shown the largest growth in the 1P data for fields, the phenomenon applies to many regions around the world. Even in the UK, with its relatively good data, the proved oil reserves total has for many years stood (and indeed, still stands) at only roughly half the quantity in the industry proved-plus-probable ('2P') reserves, see Table A2.1.

This would seem rather a lot of detailed information, but in fact the main conclusion is quite simple. Since UK government 'Brown Book' data have been used in the BP *Stats.* data, proved reserves have been low numbers, and changed little from year to year. When earlier the *Oil & Gas Journal* data were used for these data, UK reserves were considerably higher; and indeed were probably close to the proved-plus-probable values. As part B of the Table shows, over at least the last two decades the UK 2P reserves have stood at about twice the size of the proved (1P) reserves.

The full reason that the UK's proved reserves have been so much below the 2P reserves still needs elucidating. It almost certainly reflects, in part, reserves reported by oil companies under US Securities & Exchange Commission rules; but probably also the non-inclusion of reserves of discovered fields until the latter were sanctioned by government for development.

As mentioned earlier, the long period of static values for the UK proved reserves ratio—staying at the equivalent of roughly 5 year's supply for a quarter of a century, from 1985 to 2010—would not have mattered, except that it fooled many analysts into thinking that something special was going on. Year after year oil was

Table A2.1A Evolution of data on UK Reserves. "PUBLISHED PROVED" RESERVES: UK GOVT. DEFINITION OF PROVED RESERVES FROM 1986

Year	Gb	Year	Gb	Year	Gb
1975	16.0	1991	4.0	2007	3.6
1976	16.8	1992	4.1	2008	3.1
1977	19.0	1993	4.6	2009	2.8
1978	16.0	1994	4.5	2010	2.8
1979	15.4	1995	4.3	2011	2.8
1980	14.8	1996	4.5	2012	3.0
1981	14.8	1997	5.0	2013	3.0
1982	13.9	1998	5.2	2014	3.0
1983	13.2	1999	5.2		
1984	13.6	2000	5.0		
1985	13.0	2001	4.9		
1986	5.3	2002	4.7		
1987	5.2	2003	4.5		
1988	4.3	2004	4.5		
1989	3.8	2005	4.0		
1990	3.8	2006	3.6		

Notes

– Data are year-end reserves for the years shown

– Data prior to 1983 are presumably from the corresponding year-end issues of *Oil and Gas Journal* ('*O&GJ*')

– Data 1983–1985 are from BP *Statistical Review* of the years shown plus 1, where the source is given as the corresponding year-end issues of *The Oil & Gas Journal*; and the category is given as: "Published Proved" Oil Reserves

– Data since 1986 (but not prior) roughly agree with the BP *Stats. Review* Excel spreadsheets (various dates), where data are listed since 1980. In The BP *Stats. Review* 1987 printed issue it says the bulk of the reserves data are still *O&GJ* data, except now the UK reserves specifically are from the UK DTI *Brown Book*

Table A2.1B Evolution of data on UK Reserves. Proved-plus-probable reserves ('2P')

	Year	Gb
USGS	1996	9.7
C/U	2005	9.3
C2013	2010	6.6

Notes

– USGS: Data from USGS year-2000 Assessment; original data from Petroconsultants database

– C/U: Campbell/University of Uppsala estimate for 'Regular conventional' oil

– C2013: Campbell (2013) estimate, again for 'Regular conventional' oil

being produced, but thc proved reserves were not falling. This annual replacement of the proved reserves was thus very widely ascribed, including within the oil industry, the UK government and the IEA, as being primarily due to improvements in technology; horizontal drilling and 4-D seismic being frequently cited. The real explanation of course was simply that as the *proved* reserves were produced, reserves in the *probable* category became re-classed as proved.

But why did analysts not see this for what it was? The reason lies partly in the misleading definition of proved reserves given earlier: those quantities that "with reasonable certainty can be recovered in future under existing economic and operating conditions". Most analysts then—and still today—treat proved reserves as a fairly accurate measure of the amount of oil likely to be available.

Figure A2.1, though a little complex to read, sets all this out plainly. It shows UK data on cumulative production, and for 1P and 2P reserves; and, importantly, also shows the estimates made at the time of the total amount of oil likely to be recovered in UK waters (the 'ultimate'). The data are taken from issues of the UK's *Brown Book* for the years indicated.

Look initially at the data for 1974. Offshore production had not started, so the cumulative production was essentially zero. Proved reserves from fields discovered at that date (see Fig. 2.9 left) were being reported under the original 'close to 2P' rules, so were fairly significant; on top of which were probable reserves, and then

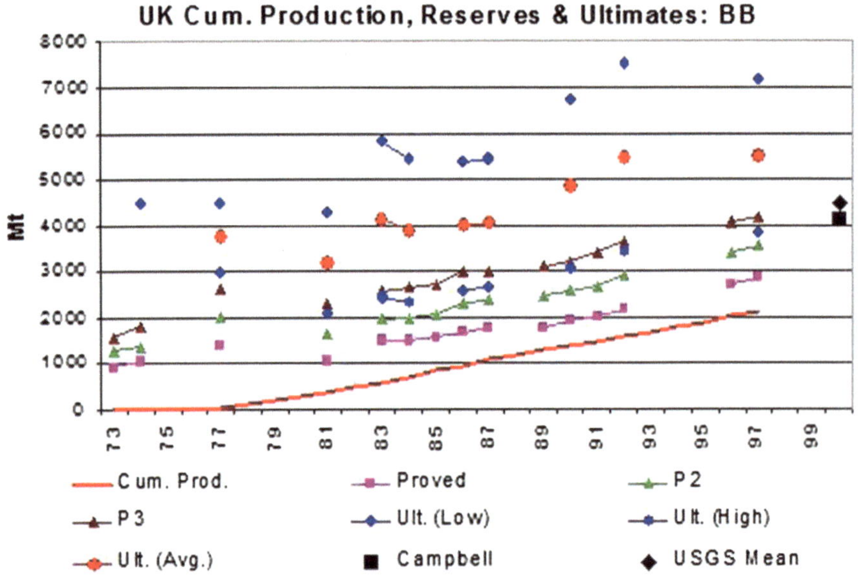

Fig. A2.1 UK oil data from the UK government's *Brown Books* ('BB') at the dates indicated, over the period 1973–1997.

Notes

- Cum. Prod.: Cumulative production.
- Reserves: Proved; P2: Proved-plus-probable; P3: Proved-plus-probable-plus-possible.
- Ult. (Low, High, Avg.): URR estimates given in the '*Brown Book*'; low, high; and average of these two values.
- Also to *right* of graph: URR estimates from Campbell (for 'Regular conventional' oil); and USGS year-2000 estimate (ex-NGLs, no reserve growth allowance).

See text for discussion

possible. At the same date, the government gave a single estimate for the likely total recoverable, which stood at 4500 million tonnes (33 Gb).

By 1977 more fields had been discovered, so the 'old-basis' proved reserves had grown significantly, and likewise the 2P and the 'proved plus probable plus possible' ('3P') reserves. By this point, the government gave a range for the ultimate, and the plot shows the low value, the high value and the average. As can be seen, the average (red dot) had fallen a bit from the 1974 estimate.

By 1987, the lower rule for reporting the proved reserves had been adopted; but despite this, the average value for the ultimate (at about 30 Gb) was little changed from the 1974 estimate.

In the subsequent years, the average value of the 'ultimate' grew somewhat, to just over 40 Gb. But as the numbers at the right of the plot indicate, and see Fig. 2.9 (right) and Fig. 2.11, estimates by others clustered between 29 and 34 Gb; all very close to the original 1974 estimate of 33 Gb.

Perhaps the most striking lesson from this plot is how easy it has been with the information displayed to make a reasonable guess for the date of the UK peak. For example, a simple fit to production of an isosceles triangle of area equal to the apparent asymptote of discovery has long given a reasonable indication of the *date* of peak (though naturally overstates production at peak, as such a triangle does not reflect the region's long decline curve). An equivalent calculation results if the original 1974 estimate for the UK's ultimate of 4500 Mt (33 Gb) is combined with the 'mid-point peaking' rule. On this basis, the UK's resource-limited peak would be expected when cumulative production reached 2250 Mt. This was not in 1984, the first apparent peak, as by then cumulative production had reached only 730 Mt, but in 1997. Given the general straightness of the cumulative production line, despite the trough from 1985–1995, this date could be (and was) predicted with reasonable precision from the outset of production. Piece of cake, really.

There remains, however, the issue of real 'reserves growth' in the UK data that in the 2P data caused by technology and price, and this is discussed in Annex 4.

(iii) *Regions generally*

Above we have looked at the specific cases of 1P reserves being low estimates of the 'true' 2P reserves for Canada, the US and the UK. But for all regions generally, we expect this to be true.

First, 1P estimates, certainly when reported by commercial oil companies are reported under SEC rules, and these (until recently, when 2P reporting has also been permitted) were always conservative values.

Second, proved reserves *in regions*, if generated by arithmetic addition of field proved reserves (as is often the case), will be underestimates for a statistical reason. Adding data of a given probability does not lead to an estimate of the same probability, except for 50 %-probable data. In the case of high probabilities (for example, where proved reserves are judged as having a 90 % chance of being exceeded), such arithmetic addition significantly underestimates the total at the same probability level.

Over time, however, as fields get produced one would generally expect the difference between the proved estimates of the reserves of either a field or region, and the estimates based on proved-plus-probable reserves, to diminish, as the conservative proved reserves data grow to become close to 2P estimates. This indeed is now the case for many—but by no means all—oil-producing countries. Thus the significance of the understatement of proved reserves is no longer the issue it once was, though it still needs to be taken into account. Its main significance lies in the misleading conclusions so many analysts have drawn about the nature of 'reserves replacement', as explained above and in Chap. 5

A2.1.2 Overstatement

Having looked at the understatement of proved reserves in some detail, we now look at the second problem mentioned above, that of overstatement.

Though most countries' public-domain proved (1P) oil reserves are, as one would expect, smaller than the industry 2P data, in some anomalous but very important cases the reverse is true, and the country's published 1P reserves significantly *exceed* the industry data 2P reserves. These cases are mainly the result of the OPEC 'quota wars' step-changes in proved reserves that have been written about extensively, and which may constitute an over-reporting, above the correct 2P reserves, by as much as $\sim$300 Gb.

Table A2.2 shows the large 'quota wars' step-changes that occurred in the proved reserves data for six OPEC countries in the 1980s; and also the subsequent long periods of no change in some of these data. Also shown are the countries' annual production in 2014 to illustrate the magnitude of change one might have expected in the reserves data if there were no major new discoveries in the years concerned.

The case has been made that some of the 'step-changes' shown in Table A2.2 reflect the countries' newly nationalised industries reporting 'true' reserves, rather than more conservative values bequeathed them by the oil majors. There may well be a degree of truth to this, but this is a far from sufficient explanation of both the size and dates of these changes. Campbell's reading of these data (personal communication) is as follows:

> *I think that when Kuwait increased its reported reserves in 1984 it did so by reporting original, not remaining, reserves by not deducting past production. In fact this is industry practice for defining the relative ownership of a field that crosses a lease or national boundary, and it made sense for OPEC to use at a basis for setting quotas. [Then] in 1986 Kuwait announced a small possibly genuine increase, but this probably proved too much for the other OPEC countries that were competing for quota at a time of low oil prices as they desperately needed the revenue on which they had come to depend. In 1986 Abu Dhabi matched Kuwait, Iran went one better at 93 Gb, and Iraq capped both a year later at a rounded 100 Gb.*

Table A2.2 'Quota Wars' step-changes in some OPEC proved reserves; and subsequent 'static' reserves data

	UAE	Iran	Iraq	Kuwait	Saudi Arabia	Venezuela	
1980	30.4	58.3	30.0	67.9	168.0	19.5	
1981	32.2	57.0	32.0	67.7	167.9	19.9	
1982	32.4	56.1	59.0	67.2	165.5	24.9	
1983	32.3	55.3	65.0	67.0	168.8	25.9	
1984	32.5	58.9	65.0	92.7	171.7	28.0	
1985	33.0	59.0	65.0	92.5	171.5	54.5	
1986	97.2	92.9	72.0	94.5	169.7	55.5	
1987	98.1	92.9	100.0	94.5	169.6	58.1	
1988	98.1	92.9	100.0	94.5	255.0	58.5	
1989	98.1	92.9	100.0	97.1	260.0	59.0	
1990	98.1	92.8	100.0	97.0	260.3	60.1	
1991	98.1	92.9	100.0	96.5	260.9	62.6	
1992	98.1	92.9	100.0	96.5	261.2	63.3	
1993	98.1	92.9	100.0	96.5	261.4	64.4	
1994	98.1	94.3	100.0	96.5	261.4	64.9	
1995	98.1	93.7	100.0	96.5	261.5	66.3	
1996	97.8	92.6	112.0	96.5	261.4	72.7	
1997	97.8	92.6	112.5	96.5	261.5	74.9	
1998	97.8	93.7	112.5	96.5	261.5	76.1	
1999	97.8	93.1	112.5	96.5	262.8	76.8	
2000	97.8	99.5	112.5	96.5	262.8	76.8	
2001	97.8	99.1	115.0	96.5	262.7	77.7	
2002	97.8	130.7	115.0	96.5	262.8	77.3	
2003	97.8	133.3	115.0	99.0	262.7	77.2	
2004	97.8	132.7	115.0	101.5	264.3	79.7	
2005	97.8	137.5	115.0	101.5	264.2	80.0	
2006	97.8	138.4	115.0	101.5	264.3	87.3	
2007	97.8	138.2	115.0	101.5	264.2	99.4	
2008	97.8	137.6	115.0	101.5	264.1	172.3	
2009	97.8	137.0	115.0	101.5	264.6	211.2	
2010	97.8	151.2	115.0	101.5	264.5	296.5	
2011	**97.8**	**151.2**	**143.1**	**101.5**	**265.4**	**296.5**	
2012	97.8	157.0	150.0	101.5	265.9	297.6	
2013	97.8	157.0	150.0	101.5	265.9	298.3	
2014	97.8	157.8	150.0	101.5	267.0	298.3	(Gb)
Ann. prodn.:	1.4	1.3	1.2	1.1	4.2	1.0	(Gb/yr. in 2014)

Notes The earlier data highlighted indicate the large step-changes in proved reserves from 1982 to 1988 for the countries shown. These occurred at a time when the oil price had fallen sharply, and one factor in a country's OPEC quota allowance was its declared proved reserves. Note that no major discoveries occurred at the dates highlighted; though Venezuela might legitimately argue that changes in extraction technology allowed more Orinoco oil to be 'proved-up'

Also highlighted are more recent cases of possible 'quota manoeuvring' since 2008
Also shown are the very long sequences of 'static' reserves (i.e., reserves that did not change over a number of years, despite significant levels of production having occurred)
Source D. Freedman, drawn from the BP *Stats. Review*; new data added since 2012. (Data may not be as of the date shown, but reflect that in one or more issues of the *Stats. Review*.)

Other sources also have pointed to probable problems with these OPEC reserves data. For example, the Wikipedia article: 'Oil Reserves' (accessed 28th March 2015, and here slightly edited) says:

- "Sadad al-Huseini, former Head of Exploration and Production at Saudi Aramco, estimates 300 Gb of the world's 1200 Gb of proven reserves should be recategorized as speculative resources, though he did not specify which countries had inflated their reserves. (Source: "Oil reserves over-inflated by 300bn barrels—al-Huseini". October 30, 2007. Retrieved 2008-08-23.)
- Dr. Ali Samsam Bakhtiari, a former senior expert of the National Iranian Oil Company, has estimated that Iran, Iraq, Kuwait, Saudi Arabia and the United

Arab Emirates have overstated reserves by a combined 320–390 Gb and has said, "As for Iran, the usually accepted official 132 Gb is almost one hundred Gb over any realistic assay." (Source: "On Middle Eastern Oil Reserves". ASPO-USA's Peak Oil Review. February 20, 2006. Retrieved 2008-08-20.)

– *Petroleum Intelligence Weekly* reported that official confidential Kuwaiti documents estimate reserves of Kuwait were only 48 Gb, of which half were proven and half were possible. The combined value of proven and possible [reserves] is *half of* the official public estimate of proven reserves." (i.e., of 101.5 Gb, see Table A2.2). (Source: "Oil Reserves Accounting: The Case Of Kuwait". *Petroleum Intelligence Weekly*. January 30, 2006. Retrieved 2008-08-23.)"
  [On these Kuwait data, Campbell (personal communication) commented: "Recently the Kuwait Minister claimed 24 Gb proved and 54 Gb proved & probable; my current estimate gives Kuwait's 2P reserves as 56 Gb."]

A2.1.3 Non-statement

Finally, in the main public-domain 1P data (those for example in the BP *Stats. Review* or the *Oil & Gas Journal)*, the data on proved reserves are frequently not updated annually, and sometimes can remain static for very long periods of time.

For example, in the OPEC data in Table A2.2, while the problem of the 'step-changes' in these data is now fairly widely known, the problem of 'static' data is much less so, although both problems were described clearly in the Campbell & Laherrère *Scientific American* article (1998). As Table A2.2 shows, for example, the UAE-proved reserves have stayed effectively unchanged at close to 98 Gb since 1987, despite the country's annual production averaging roughly 1 Gb/yr. over this nearly 30-year period. This problem of 'static' proved reserves applies to many countries, in addition to those in Table A2.2; for example in the end-2014 *O&GJ* data for 106 oil-producing countries, 66 of these reported no change in proved reserves from the previous year.

For this reason also, that of non-statement of changes in their values, the 1P data are significantly in question; and more importantly as a consequence, any *trends* in these data are especially misleading.

Overall, it has been disappointing that so many analysts have relied on proved oil reserves data without sufficiently investigating, and drawing far more attention to, their clearly very questionable nature.

A2.2 Proved-Plus-Probable ('2P') Reserves Data

Now we turn to the oil industry proved-plus-probable (2P) reserves data.

As mentioned in Chap. 1, such data can be gathered for individual oil fields from a wide variety of published industry and government sources, but with considerable effort. And for fields that are clearly into their production decline phase, one can make one's own estimates of a field's total size (i.e., of its URR, comprised of cumulative production plus 2P reserves) from simply its past production data. This is done by using the linearisation technique shown in Fig. 2.3; by plotting the field's annual production vs. its cumulative production, and extrapolating the—typically—fairly straight line to the abscissa. If available, geological, reservoir and production engineering knowledge should be incorporated into this estimate where appropriate. But as Laherrère points out, surprisingly, wholly reliable public-domain field *production* data are available only for the UK, Norway and the US offshore (the latter from the Bureau of Ocean Energy Management).

As also mentioned in Chap. 1, as an alternative to the above approaches, large commercial datasets of 2P data for individual fields can be purchased. These come from firms such as IHS Energy, Wood Mackenzie, PFC Energy and Rystad Energy, and where the data have been assembled and checked; and where there is also much proprietary information. Such datasets however are generally very expensive; a licence for IHS Energy's full Exploration & Production dataset costs in the region of $100,000 annually.

(And note that Laherrère cautions that while these datasets are indeed generally '2P', i.e., proved-plus-probable estimates, and hence are the data to use for predicting future production, given the multiplicity of sources from which the data are drawn, in probability terms the data values can be P50 (i.e., 50 % probable), mean (~P45), or most-likely (mode, ~P65). As a separate caution, note that for IHS Energy data, and possibly also for that from other data suppliers, the data for the US and Canada non-frontier areas are only 1P.)

As set out in Chap. 1, simpler 2P datasets are available at moderate cost. Certain by-field data can be purchased from Globalshift; and probably from Rystad Energy; and possibly Peak Oil Consulting, Richard Miller and Uppsala University. In addition, extremely useful *by-country* 2P data are available from IHS Energy's 'PEPS' database, where researchers should use the version with data back to 1834. These data are affordable to any reasonably-funded academic research project, and are used in a number of the Figures given in this book. Annex 3, below, indicates ways in which this dataset can be used.

Some collected 2P data are also now in the public domain. For the 'all-oil' global data, of both discovery and production, see the charts of IHS Energy data in Miller and Sorrell (2014), summarised here in Fig. 2.23. For conventional oil, adjusted data both globally and for a number of countries are given by Jean Laherrère on websites and elsewhere (including as charts in *The Oil Age,* January 2015).

Charts for Campbell's adjusted 'Regular conventional' oil for a wide range of countries, and including essays on the associated petroleum geology, are in his excellent *Atlas of Oil & Gas Depletion* (2013). The latter covers all-oils, but the main by-country data are for 'Regular conventional' oil. The key graphs are the 'Status of Oil Depletion' plots. These give, by country, the critical cumulative backdated 2P discovery data, to show how much oil has been found, the country's cumulative production, and also Campbell's judgement on the 'extrapolated-discovery' URR. The evolution over time of that country's 2P reserves, and yet-to-find, can be examined from the differences between these data.

Useful plots of past and forecast production (but not of the crucial oil *discovery* information) for all oil-producing countries, based on detailed by-field 2P data, are free on the Globalshift website (www.globalshift.co.uk). In addition, Rystad Energy's UCubeFree facility gives past and future production by-country (but again, not the discovery data), based on their estimates of 2P data.

In terms of understanding the above data, it is important to note that 2P data from almost any source (and certainly the IHS Energy 2P data) are *backdated* values, i.e., *today's* estimates of what any given field or region contains. For the significance of this see Laherrère's (2011) presentation: *Backdating is the key*. It comes about simply because the data providers wish to keep their data current, and if new data comes forward on a change in the estimated original volume of oil to be extracted from given field, this new information is logged against the field, which maintains its original discovery date.

A2.3 Comparison of 1P Data Versus 2P Data

As indicated earlier, a crucial part of the difference between global 1P and 2P oil reserves data has been their very different evolution over time, which in turn has led to analysts drawing very different conclusions about the future availability of oil. Here we give four examples of comparing 1P oil data with 2P data: for the US, for the UK, for some OPEC countries and at a global level.

A2.3.1 US Data

We start with the US reserves data, both 1P and 2P, as it has been these data that underlay many of the arguments used in the peak oil debate—both for and against the case for 'peak oil'. Understanding the difference in these data helps clarify why the controversy occurred. Four plots of US oil discovery (or of 'apparent discovery', where the 1P data are used) are given in Fig. A2.2.

All the plots in Fig. A2.2 are on a cumulative basis (i.e., give summed values). In three of the plots (those on the left, plus the one top right), cumulative discovery is compared to cumulative production, and are on a cumulative basis versus year; i.e. show the total amount of oil discovered and produced by the year indicated. The fourth plot (lower right) also gives US cumulative discovery, but here versus the number of exploration (new-field wildcat) wells drilled (i.e., is a so-called 'creaming-curve' plot).

Where the data for cumulative discovery are derived from 2P data, cumulative discovery is simply the summation of the estimated backdated URR of all the fields discovered by the date indicated (or correspondingly, in the fourth plot, by the number of new-field wildcats drilled). Where the data are for cumulative discovery

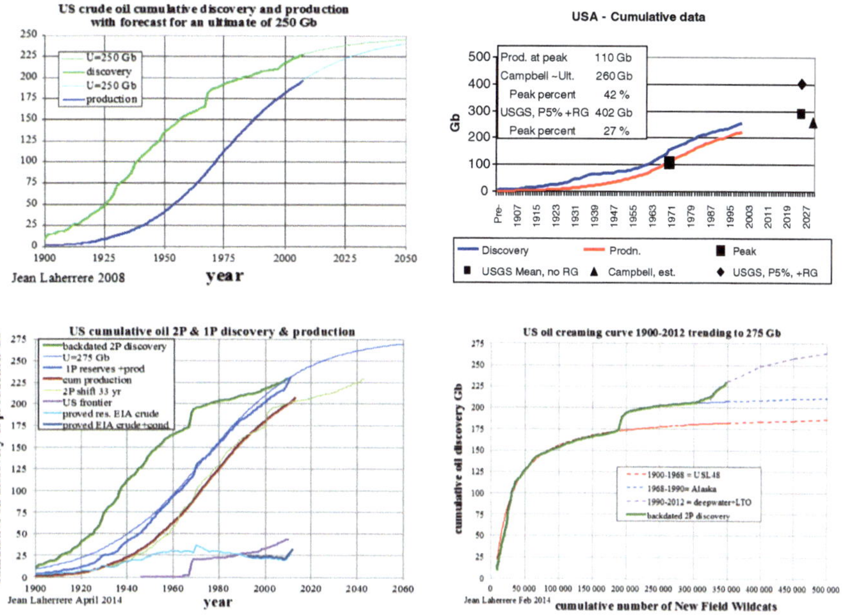

Fig. A2.2 US Oil Discovery Data: Contrasting 1P data with 2P.

Top left Simple plot of cumulative proved-plus-probable ('2P') US discovery data, and production, 1900–2007. *Source* Laherrère, from US DOE (1990) and Attanasi and Root (1994). Data exclude NGLs. (By the year 2000, cumulative production of NGLs adds, very approx., ~35 Gb.).

Top right Plot of cumulative proved ('1P') US discovery data, and production, 1900–2000. *Source* IHS Energy (Petroconsultants) 'PEPS' database, year-2000, with permission. Data include NGLs; but recall that 'PEPS' data for the US and Canada (except for frontier areas) are 1P, not 2P. [Also shown on this plot is the US peak date for its conventional oil production; and also the Campbell and USGS estimates, as of year-2000, for the all-US (not just Lower-48) conventional oil ultimate (URR).].

Lower left A more recent, rather more complex plot from Laherrère. As of ~2012 US '1P' discovery (cumulative production plus 1P (proved) reserves) had grown to the same value as backdated 2P discovery; i.e. US 1P reserves have now reached ~2P (~50 %-probable) values.

Lower right Cumulative US 2P discovery as a 'creaming curve' (i.e. vs. cumulative number of new-field wildcats, NFWs). This illustrates of one of the methods used by Campbell and Laherrère in the Petroconsultants report (1995) to estimate URR for a region, by 'extrapolation by eye' of the discovery creaming curve out to roughly double the number of NFWs. As the plot shows, this method yields a URR of ~185 Gb for US Lower-48 ('USL48'), in agreement with Hubbert's several methods to estimate this quantity; of ~210 Gb if Alaska is included; and a current value of ~275 Gb if now deepwater Gulf of Mexico and recent light-tight oil ('LTO') are included. (These 'creaming curve' estimates for the US URR should be contrasted to the historical USGS estimates post-1971, and for some years after, of US URR values up to ~600 Gb.)

using 1P data (plot on top right), cumulative discovery is calculated by the summation of cumulative production to the date indicated plus the proved reserves at that date.

The top left chart of the figure shows one of the few datasets to cover US discoveries based on proved-plus-probable (2P) data for field discovery size, while in the top right chart shows the apparent quantity of oil discovered if only the proved reserves (1P) data plus cumulative production are used to indicate discovery.

The difference between these datasets is dramatic. While both sets of cumulative data (that from the 2P data, and that from the 1P) trend towards similar asymptotes for 'ultimate' discovery, the timing of the discoveries is very different. In the top right chart, the 1P data support the 'economic' view that 'oil reserves are just inventory', with US 1P cumulative discoveries staying (except for a brief period in the 1930s) only just ahead of cumulative production for over a century; and resulting in the well-known characteristic of the US R/P ratio of proved reserves to production being an apparent fixture at 10 years' of production.

By contrast, the US discovery data using the backdated 2P field size data show that cumulative oil discovery, rather than staying just ahead of cumulative production, surged ahead of production from 1900 up until about 1950, with only then cumulative production starting to eat into the large cushion of oil that had been discovered.

It is true that reserves growth in the *2P data*, due to advances in recovery technology etc., is not factored into this picture, (and see the discussion of this in Annex 4); but lack of non-backdated (i.e., current) 2P data unfortunately makes such a comparison difficult or impossible.

But despite this, the left chart based on the 2P data clearly reflects much of the reality of US oil discovery. The inflection in the discovery curve on this chart shows the peak of US Lower-48 discovery as occurring in the mid-1930s. This followed the discovery of the East Texas field and other large Texas fields, with US discoveries then starting to tail off, a fact that began to be recognised by about the late-1940s (see Yergin 1990). And it was this falling discovery trend that had allowed realistic estimates to be made by various oil industry experts of the size of the US Lower-48 'ultimate' (ex-NGLs) of 150 and 200 Gb; data that were then used by Hubbert in his 1956 forecast for the US production peak to occur between 1965 and 1970 based on this URR range.

This '2P' chart also clearly shows the discovery of Prudhoe Bay in Alaska in 1968, as well as the more recent discovery gains coming mostly the from deep offshore in the Gulf of Mexico plus the recent 'light-tight' oil. Overall, these charts —which explain for the US 'what oil was found when', and also how the country's 1P versus 2P reserves have evolved—help clarify some of the confusion that underpins the 'peak oil' controversy.

A2.3.2 UK Data

Similar lessons can be learned by looking at the corresponding data for the UK. A graph comparing the evolution of UK 1P reserves data with 2P is given as Fig. 10 of Sorrell and Speirs (2014); and see also Fig. A2.1, above. As with the US data, the difference between these two datasets is large.

A2.3.3 Some OPEC Countries

Now we compare 1P oil reserves data with 2P data for some Middle East OPEC countries. This is done in Table A2.3, which compares some of the proved (1P) reserves data given in Table A2.2 with data from two industry sources for proved-plus-probable ('2P') reserves.

As has been mentioned earlier, against intuition, the Middle East 1P data for the countries shown are considerably *larger* than either of the 2P estimates. The geopolitical consequence of this discrepancy becoming apparent in terms of future production from these countries is likely to be significant.

Table A2.3 Middle East Remaining Reserves (Gb)

	Proved (*BP Stats.*) ('1P')	Campbell[b] ('2P')	Petcon.[c] ('2P')
UAE	97.8	47	57
Kuwait	101.5	56	55
Iraq	150.0[a]	66	96
Iran	157.8	68	101
Saudi Arabia	267.0	164	223
Total (rounded)	**775**	**400**	**530**

Notes

'1P': Proved reserves, *BP Stats. Rev. 2015*, includes NGLs

'2P': Proved-plus-probable reserves (see specifics below). These 2P data are not directly comparable due to different definitions and source dates

[a]Iraq is something of a special case, having recently opened up its fields to outside cooperation, as the following quotation illustrates:

"The new reserve figure was reached with the help of international oil companies working at 12 fields in Iraq". Hussain al-Shahristani, the oil minister, said in a news conference. "The rise in reserves from West Qurna and Zubair has considerably contributed to the [overall] rise; West Qurna has reserves of 43.3 bn barrels, making it the second largest field in the world."

[b]Campbell: 'Estimated future production from known fields', Regular conventional oil, ex-NGLs; 2014 data (personal communication); see also Campbell (2013) for his 2010 estimates

[c]IHS Energy; year-2000 data, 'all-oil', includes NGLs. [Note, some more recent IHS Energy data show higher values; the important topic of recent changes in IHS 2P reserves data for certain Middle East OPEC countries is discussed in connection with Fig. 2.20, and in Section A4.1, below.]

A2.3.4 Global Data

Now we turn to global data on oil reserves. Figure A2.3 shows the evolution over time of the difference between the public-domain data for global 'all-oil' proved (1P) reserves, and the view of one experienced analyst, Jean Laherrère, of the corresponding data for the likely global backdated proved-plus-probable (2P) reserves of conventional oil. The latter are based on interpretation and adjustment of oil industry 2P oil data from a variety of sources.

The main lessons from Fig. A2.3 are as follows:

1. If 1P data are used, global reserves of 'all-oil' (including condensate and NGLs) show an apparent very encouraging ever-upward trend. (Note that the OPEC 'quota-wars' gains in these 1P reserves are clearly visible, as are the majority of the increases in these reserves since the year 2000 that are due to the inclusion of tar sands and Orinoco reserves; and the smaller amount, since about 2010, from the inclusion of US 'light-tight' oil reserves.)
2. By stark contrast, the size of the global backdated 2P global *conventional oil* reserves (i.e., after the very heavy oils, light-tight oil, & NGLs have been taken out) peaked at about 1980, and have been in steady decline since.
3. So the question immediately arises, why is there such a dramatic difference in these two measures of global oil reserves? In analysing this difference note that up until about the year 2000 both measures of reserves covered roughly the same class of oil (i.e., only conventional oil), as up to roughly about this date the quantities of tar sands, Orinoco, light-tight' oil and NGLs included in the 1P reserves data were relatively small. Instead, the main reasons that the 1P global reserves have been so far below 2P reserves for most of the period shown have been as follows:

   - The 1P data exclude the *probable* reserves that are included in the 2P data, and where, especially in the early years (and particularly for US and Canadian data), the size of these omitted probable reserves were large.
   - The 1P reserves were also under-reported because of the statistical problem mentioned above of aggregation.
   - The 1P data shown were current at the date shown (i.e., estimates made at that date); whereas the 2P data are backdated data, i.e., reflect 2014 knowledge of the size of remaining reserves in fields discovered as of the dates shown. (Unfortunately, it is not possible to compare 'like with like' here, i.e. current-basis 2P data with current-basis 1P, because the former data are not available, see notes earlier, even though it would be helpful if these data did exist.)

4. Over much of the period shown the difference between the two classes of reserves, 1P and 2P, has been very large; for example, in the mid-1960s 2P reserves were larger than 1P by over 500 Gb. Later, as fields aged, proved reserves moved towards proved-plus-probable values; but, separately, the 1P

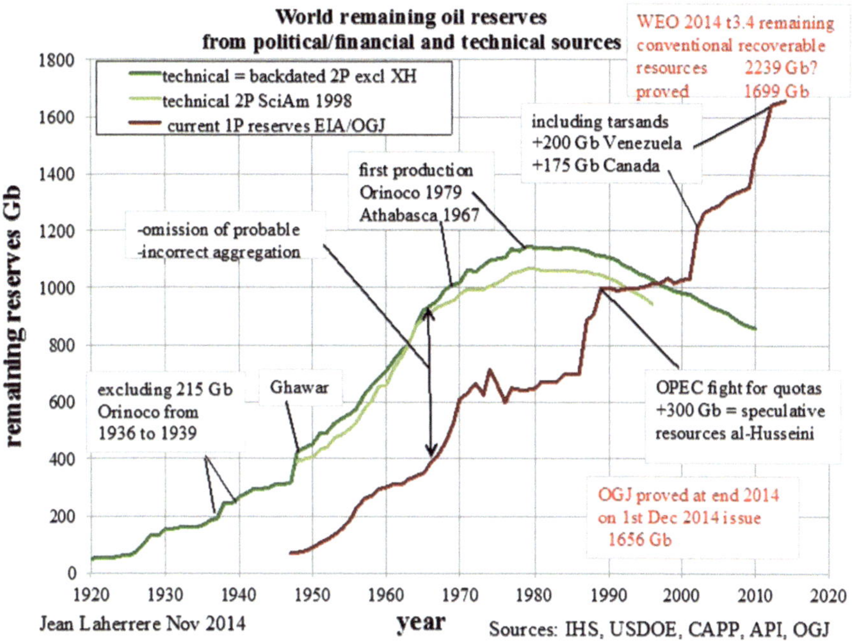

Fig. A2.3 Difference between global backdated proved-plus-probable ('2P') oil reserves as estimated by Jean Laherrère, based on adjusted oil industry 'scout' 2P data, and current-basis proved ('1P') global oil reserves as given by public-domain data.

Leftmost (*dark green*) line (exhibiting a peak in ~1980 at ~1150 Gb): Laherrère's estimate of global 2P backdated oil reserves, 1920–2010, excluding extra-heavy oils (primarily tar sands and Orinoco oil). Data are from industry databases but adjusted by removing 300 Gb of almost certainly "political" reserves from Middle East reserves, and 30 % of FSU reserves.

Next leftmost (*grey-green*) line: The corresponding data as used in the 1995 Petroconsultants study that underpinned the Campbell and Laherrère 1998 Scientific American '*The End of Cheap Oil*' article.

Rightmost (*magenta*) line: Data from the US EIA and *Oil and Gas Journal* (OGJ) of public-domain current-basis global 1P oil reserves, 1947–2014, *including* extra-heavy oil.

Comments: Estimated backdated 2P global oil reserves, excluding extra-heavy oil, peaked in 1980. By contrast, global 1P oil reserves, also excluding extra-heavy oil, which were only included later, rose consistently to ~1997, to roughly match the 2P data (implying that at that date OPEC 1P overstatements roughly matched the industry global probable reserves); and the increased subsequently, in part due to adding in Canada and Venezuela extra-heavies

data came to include increasing amounts of non-conventional oil (excluded from the 2P data in this plot).

5. The rough match between the 1P and 2P oil reserves data around 1997 implies that the OPEC 1P overstatements probably totalled about the same at that date as the industry global conventional oil probable (only) reserves.
6. For comparison of Laherrère's judgement of the evolution of global 2P reserves data, as shown here, with IHS Energy's 'all-oil' 2P reserves data (which include NGLs and non-conventional oils) see Fig. 2.23. Although the latter 2P data are

different from those shown here (due to inclusion of non-conventional oils, and to not including the adjustments made by Laherrère), the same critical decline of global 2P reserves since the 1980s, based on new fields and projects, is apparent.

Overall, the main conclusion to draw from Fig. A2.3 is that for largely conventional oil, the size of global 1P reserves were on a steady ever-upward trend until 2002; and have since increased more sharply still due to inclusion of the proved reserves for tar sands and Orinoco oil. By contrast the industry data show that the global 2P reserves of conventional oil peaked in about 1980 and have been in steady decline since.

Annex 3
Notes on Use of IHS Energy 'PEPS' By-Country Data

As mentioned earlier, the full by-field global dataset from IHS Energy is very expensive, but the company's 'PEPS' by-country dataset is available at a price that funded academic research can afford, and provides quantitative answers to many of the issues discussed in this book. The dataset contains information by country since 1834 including, *inter alia*, annual discovery and production, and hence, by subtraction, reserves. The discovery data (and hence reserves also) are backdated 'proved-plus-probable' values, summed for each country from the individual field 2P data; except for US and Canadian non-frontier areas, where the data are only proved values.

The data have to be kept confidential in general, but IHS is happy to discuss release of data if used for academic projects, provided adequate safeguards are adhered to. If within a research project, it is possible to agree a protocol that allows the researcher to use the database, but keep detailed results from the public domain, and marking any final report lodged with the academic institution as confidential. Papers intended for publication have to be agreed by IHS, but release of specific data can often be achieved to the satisfaction of both the academic and the company; for example, where the data are aggregated or otherwise disguised, such as by removal of axis labelling on charts.

For small research projects, the following investigations are possible (but other approaches will undoubtedly occur to researchers):

(a) Plotting up 'nine-graphs-per-page' graphs, or similar. These graphs are extraordinarily useful for understanding peak oil.
    The nine graphs for each country of interest are by-date on the abscissa, from 1900 to the present day, and each shows 2P discovery and production. These can be set out in a square format of one page per country, and arranged as:

    - by column: onshore oil; offshore oil; total oil;
    - by row:
        - annual data (i.e., annual discovery as a vertical bar, & annual production as a line);
        - 5-year smoothed discovery, and annual production; both as lines;
        - cumulative discovery and cumulative production; as lines.

R.W. Bentley, *Introduction to Peak Oil*, Lecture Notes in Energy 34,
DOI 10.1007/978-3-319-26372-4

It is essential that these graphs have only one ordinate scale (as in Ivanhoe's original world graph in the *Oil & Gas Journal*, and Campbell's comparable graphs in the *Atlas*), as this lets the user see directly the transcendental connection between discovery and production: 'if it has not been found it cannot be produced'. It is this need for a single ordinate scale that makes the requirement for the '5-year smoothed 2P discovery' line, as in some important countries annual production is too small to see usefully versus annual discovery.

The first row 'annual data' graphs is very telling, letting the researcher see at a glance what has been found and when; and when this oil was produced. Nearly every country has a fascinating story to tell; for example:

- The US with its finds pre-1900; its big finds in the 1930s, its largest and much later Prudhoe Bay find and, more recently, the offshore story. (But caution, as the discovery data here are 1P, not 2P).
- The UK with its slow small onshore finds, then burst of major offshore finds that let the date of peak become clear just five years after their discovery started, and when offshore production was only just beginning; Fig. 2.9.
- The graph for Colombia—the question of access across the mountains; Campbell having identified the prospect 30 years before.
- The Indonesia graph—a history of onshore and offshore discoveries, and hence why she was forced to leave OPEC; Fig. 2.15.
- Russia—the effects on both discovery and production of the end of the Soviet Union; and now at conventional plateau.
- Nigeria—unfortunately, peak in sight for that troubled country.
- Iraq—major finds, but a dramatic story on production.
- Iran—major finds, but with the peak in sight.
- Saudi Arabia—fallow fields, but no substantive recent finds; Fig. 2.17.

The second row of the 'nine graphs', of 'smoothed-discovery' plots, lets the link between the amount discovered and that produced so far become clearer, as given by the areas under the two curves.

The third row, of cumulative graphs, are perhaps the most useful, as the researcher can see if cumulative production exceeds about 60 % of cumulative discovery, in which case the PFC rule says that peak is likely (and indeed, the UKERC study indicated that when peak occurred this percentage was often actually less).

The same cumulative graphs, for the many countries where cumulative 2P discovery has visibly slowed, can also let the researcher estimate the date of peak, by seeing when cumulative production will exceed about 50 % of the 'discovery-extrapolated' ultimate, the latter obtained by extrapolating by eye cumulative discovery some years into the future.

In terms of work required, producing a useful set of such graphs by country, and drawing initial conclusions, could probably fit within a final-year undergraduate project (or be split between two); or else, with additional analysis, constitute the project of a taught Master's course.

Note that the results from these sort of plot are not quite as clear-cut as was the case for the plots included earlier in this book which used 'PEPS' data up to the year 2000. This is because now an increasing proportion of the 'PEPS' 'Liquids' volumes, both for discovery and production, include non-conventional oils, specifically tar sands, Orinoco oil and shale ('light-tight') oil, as well as the NGLs. Though these are not so very large in total, to see the picture for conventional oil only these volumes of non-conventional oils need to be subtracted out.

(b) Adding estimates of 'ultimate' to the by-country cumulative data plots produced above

It is very instructive to add various estimates of 'ultimate' to the above by-country cumulative data plots (shown on the plots as dots at an arbitrary future date), as in most cases extrapolating the cumulative discovery by eye shows visually how realistic these ultimate estimates are likely to be, at least in any reasonable timeframe. For 'ultimates' the researcher can use:

- the USGS year-2000 estimates, which are given by country (but without reserves growth);
- the same estimates with added reserves growth (as a lumped ratio, as was done in the UKERC report);
- more recent USGS estimates;
- Campbell's data (from the recent Campbell '*Atlas*' or his latest data by contacting him), but remembering to add-in, on top of his 'Regular conventional oil produced to 2100' (i.e., ~ultimates), allowance for deepwater, polar and heavy, where these allowances apply to only a limited number of countries;
- ultimates (or equivalently, forecast cumulative production to some distant date) from other sources, such as the IEA, EIA, Globalshift or Rystad Energy.

(c) Adding apparent reserves growth to the graphs

The UKERC report looked at by-country apparent reserves growth for some countries, as determined by the difference between cumulative 2P discovery data in the ODAC/Reading year-2000 PEPS and the same data in year-2008 UKERC PEPS dataset. It would be very instructive to illustrate this analysis by showing one or more earlier '2P discovery' lines to by-country cumulative plots (i.e., the third row of the 'nine-plots-per-page' graphs).

The researcher would then look for:

- which countries display significant such apparent reserves growth (probably only a few countries, as the UKERC report indicated);
- whether—as seemed to be the case—this apparent reserves growth has gone to almost zero over the last 15 years or so;
- which countries (if any) have significant 'apparent reserves reduction'.

And, also of course, if a researcher can talk to reservoir engineers knowledgeable about the countries concerned to explain why such reserves growth has taken place, either by specific field, or by class of fields, this would constitute publishable work.

(d) Ways of showing peak

Plots which may be new, but which seem instructive and important, and which are easy to do based on the PEPS data, are:

(i) Plot with abscissa as date (from 1930 to present), and ordinate as some measure of a country's recoverable oil volume (ultimate, or cumulative production to-date, etc.); data as just horizontal lines, one for each country, starting at that country's date of peak 2P discovery, and ending at date peak of production. (Perhaps sketch this to see what is meant.)

(ii) Plot one point for each country of: (cumulative production volume at peak)/(ultimate, or grown ultimate) as a percentage on the ordinate versus some measure of a country's recoverable oil volume on the abscissa. (i.e., these are the data examined in the UKERC report, but plotted up to see if there is any sensible correlation vs. country oil volume).

A similar plot can be done, but using (cumulative production volume at the peak production date)/(cumulative 2P discovery at the same date) as a percentage on the ordinate, again as examined in the UKERC report.

Annex 4
Further Aspects of Oil Forecasting

This Annex concerns approaches and techniques that need consideration when forecasting oil supply. In particular, it examines data reliability and forecast model assumptions (including decline and depletion rates, oil in fallow fields, and what is a realistic URR value to use) in order to draw conclusions on which of current forecasts are the more likely to be correct.

A4.1 Data Reliability

Getting reliable data is the single most important aspect of oil forecasting, and is a key differentiator between current models.

As mentioned in Chap. 1, global data on field discovery date and estimated 2P volume can be assembled from mostly publically available industry and government sources by the person or group generating the forecast, and such data can also be purchased from data companies. Estimates of URR for most categories of oil can be obtained from organisations like the BGR and USGS; while for some oil categories, such as light-tight oil, recent reports from the US EIA can be used.

However, significant caveats apply to all of these data.

- For 2P field size data caution is needed, particularly for Middle East and FSU fields.

  For example, Laherrère reduces the industry 2P data total for OPEC discovery by 300 Gb to allow for the fact that since OPEC countries have been using these data, in his judgement the size of Middle East fields in the database has grown towards the declared country proved values (see Fig. 2.20; a curious case of '2P growing towards 1P'!).

  Laherrère comments that this is because some industry datasets have been "*obliged to accept OPEC field data and their total reserves*". This is a strong statement to make, which he backs up by the example of: "*Saudi Arabia evolution of ... discoveries* [in such 2P datasets] *to the present cumulative* [of] *400 Gb (100 Gb more than in 1998) in agreement with* Oil and Gas Journal *proved* [1P] *reserves plus cumulative production*".

R.W. Bentley, *Introduction to Peak Oil*, Lecture Notes in Energy 34,
DOI 10.1007/978-3-319-26372-4

As Fig. 2.20 shows, the data for Saudi Arabian 2P cumulative discovery to 1998 apparently grew by ~120 Gb between that year and 2011 despite—as the plot also shows—the number of fields included in the database up to 1998 remaining essentially unchanged. (Laherrère also draws attention to widely quoted remarks by Sadad Al-Husseini on some Middle East data reflecting 'speculative' reserves, quoted above in Annex 2.)
It is possible that the recent upward trend in USGS data has—in part—the same cause.
For similar reasons Campbell also significantly reduces the apparent Middle East 2P discovery data; generally making his adjustments by subtracting off a country's cumulative production during those periods when its 1P reserves remained static.
Given the importance of OPEC fields to the global economy, independent audits of these fields would seem essential, though admittedly unlikely.
For FSU countries, and in part based on production-trend analysis (similar to that illustrated in Fig. 2.16), Laherrère suggests that much of the apparently 2P data are in fact 'ABC1' (i.e., roughly '3P', proved-plus-probable-plus-possible), so reduces the industry database FSU field volumes by 30 %, i.e., by roughly 100 Gb.
Laherrère also removes 200 Gb from early Orinoco discoveries, judging these as being better counted within heavy oil reserves.

- However, as mentioned previously, field sizes can increase over time due to better knowledge of the field, application of new technology or a higher oil price. This effect should not be forgotten, and is discussed in Sect. A4.3, below.
- Estimates of URR, such as the USGS 'assessed' values, must also be used with caution, see Sect. A4.7.

A4.2 Modelling Different Categories of Oil

Peak oil sceptics often say: '*the tank does not care where the gasoline comes from*', meaning that as long as there are large resources of oil available, it does not matter whether conventional, 'light-tight', tar sand or kerogen oil is used. This of course is true. But it ignores a key part of the scientific approach: the disaggregation of a problem to a level where the component parts are properly understood before re-aggregating to solve the original problem. To understand the global 'oil supply question' the scientists knew (and admittedly, today, most analysts now know) that it is necessary to disaggregate oil supply down to a level where modelling correctly captures the drivers for each component's supply.

For example, as shown in the main text, it is important to look at the supply of *conventional* oil (i.e., oil in fields) as this level of detail is needed to capture the fact that aggregate supply from fields declines for physical reasons long before all the oil in the fields has been produced. And Campbell, for example, for modelling purposes further separates out NGLs and deepwater, polar and heavy and tight oil from 'Regular conventional' oil because of their different discovery and production profiles.

A4.3 How Much to Allow for 2P 'Reserves Growth'

As Chap. 5 explained, much of the misunderstanding over 'peak oil' can be traced to apparent 'reserves growth' in 1P data, resulting from a lack of understanding of what these data report.

But here we discuss reserves growth in the 2P data, where this refers to a real change over time in the estimated 'proved-plus-probable' amount of oil that will be extracted from a given oil field, or group of fields. Typically this change increases the estimated size of the original reserves, but not always, and can be give a decrease, especially later in a field or region's life.

Such 'true' reserves growth in 2P data for a field can result from new reservoirs being discovered within the field, application of a better extraction technique, a higher oil price, or simply a better estimate of field size as more knowledge about the field becomes available. But some of apparent 2P growth results from what Campbell calls the 'U-shaped' reporting curve, where geological estimates of field size are reduced when initial engineering-based estimates are publically announced, with the latter growing back to close to the original geological estimate as the subsequent extraction methods and infrastructure are installed over time. This was the case, for example, with 2P estimates for Prudhoe Bay, as confirmed by Jeremy Gilbert, former Chief Petroleum Engineer for BP in the US.

Today the global volume-weighted recovery factor for conventional oil is probably only somewhere between ~35 and ~40 % (there being quite a lot of uncertainty on this value), implying, roughly, for a URR of say 2500 Gb, an original global conventional oil-in-place of ~7000 Gb. Hence a 1 % increase in recovery factor would yield 70 Gb, about twice the global annual rate of production. The key questions therefore are: by how much is the global recovery factor increasing annually; how much will it increase further as a result of consistent high oil prices; and when in the life of individual fields does this increase in production occur?

One approach to addressing these questions is to look at how the value for 2P discovery in a region, or globally, at a specific date in a given backdated industry dataset, has changed over time. A number of attempts have been made to do this with IHS Energy datasets, including by this author (see Campbell, (Ed.) 2011, Chap. 4). However, such results must be handled with caution as backdated data from any source will change if new fields or oil types are included in the data; if data are updated from a significantly different data source; or are updated with less reliable data.

There is a good review of reserves growth in Thompson et al. (2009) as part of the UKERC *Global Oil Depletion* study. This includes reference to the study by Klett et al. (2005) which looked at reserve growth in the IHS Energy dataset over the period 1995–2003. This found that reserve growth had apparently added 171 Gb of oil to rest-of-world 2P reserves (excluding US), i.e. an average of ~20 Gb/yr.; i.e., about two-thirds of annual consumption, and more than twice that added through new discoveries.

But a subsequent study—possibly by K. Chew—looked behind these data and found that some of this increase resulted from inclusion of new FSU data, and much of the rest was concentrated in OPEC countries, where, as Fig. 2.20 indicates, there is uncertainty as to which data—the earlier or the later—are more likely to be correct.

Thompson et al. themselves looked at more recent reserves growth in the IHS 'PEPS' dataset, and found that if US data are removed (as PEPS data are only 1P for the US non-frontier) then pre-2000 fields were estimated to have grown by 13.9 % between 2000 and 2007; i.e., about 35 Gb/yr.; roughly equal to annual consumption. But again most of this increase was in data for OPEC countries, and hence needs to be treated with caution; as does the question of whether old fields were subsequently added to the database.

More recently Sorrell and Speirs (2014) have also looked at this approach. However their analysis does not include geological knowledge, which is critical in interpreting the significance of growth curves. Nor, because of lack of access, could their study include examination of the individual field data underlying the IHS 'PEPS' database that they used. Nor—I think—did they look at the question of what fields were added to the PEPS database between their data 'snapshots'. All these checks are important to do if possible, because the central problem with the above studies is over the true interpretation of the apparent reserves growth indicated.

A4.3.1 UK Reserves Growth in UK Offshore Fields

For apparent 2P reserves growth in the UK North Sea see the relatively modest growth exhibited by the data analysed in the UKERC *Global Oil Depletion* report (Sorrell et al. 2009), and shown in Fig. A4.1.

As Fig. A4.1 shows, field growth is very variable between fields, but averaged over time, the large fields grew by about 50 %; and the smaller fields by about 25 %. These are significant increases, and should not be ignored in the modelling. But these values are less than one-tenth the US and Canadian '1P reserves becoming 2P' reserves growth values, of 600–900 %, reported above. And note that these increases are those in UK government data; and do not necessarily reflect the increases that might be seen in commercial industry 2P datasets.

A4.3.2 USGS Assessments of Scope for Reserves Growth

Over many years now the USGS has been almost unique in the world as a public institution in carrying out high quality detailed basin-based assessments of global undiscovered oil, for which they are much to be thanked. (Other institutions also

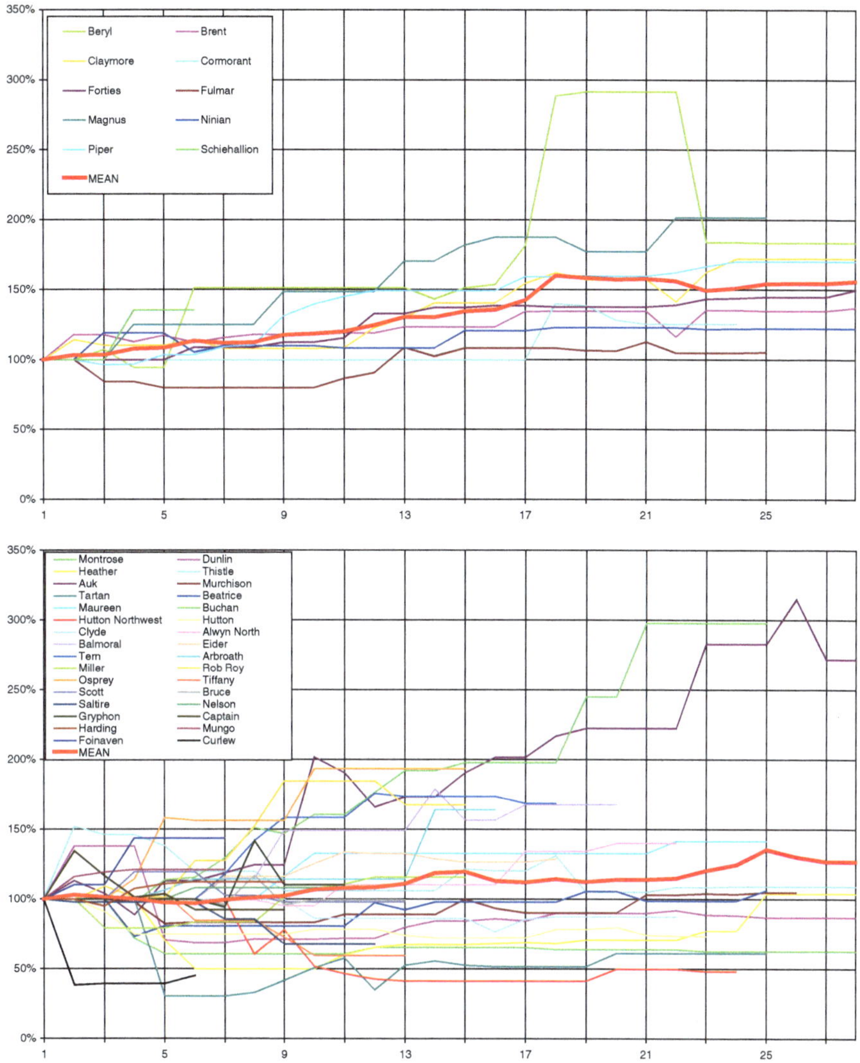

Fig. A4.1 Reserves growth for UK oil fields; '2P' data. Data from R. Miller of BP.

Upper graph UK large fields, showing the change in industry data for 'proved plus probable' (2P) reserves with time after first declaration. The Beryl field seems to be anomalous between years 18 and 22, but the trend of the data is clear: after 25 years, reserves for large fields had grown by some 50 % on average.

Lower graph UK small fields. The data are probably statistically unreliable by 25 years, as few small fields have yet operated so long. Interestingly there is no significant change in industry data for declared 2P reserves for 9 years, but then a steady growth sets in, reaching 25 % after 25 years altogether. This might suggest a very good initial estimate of field size, with only statistical fluctuation of the mean. After some 10 years, further exploration effort (driven by approaching exhaustion?) has discovered a suite of satellite fields, stacked reservoirs and other deposits entirely excluded from the initial estimates. Miller noted that "It would be interesting know whether the large fields (>500 Mb recoverable) grew from the discovery of new pools".

Table A4.1 USGS Year-2000 Assessment: world conventional oil, including NGLs

Report Date	Date of data (1 Jan.)			Y-t-F			Ultimate		
2000	1996	Cum. Prod.	Reserves	F95	Mean	F5	F95	Mean	F5
Total ex-RG		717	959	495	939	1589	2171	2615	3265
RG							265	730	1197
Total + RG							**2436**	**3345**	**4462**

Notes All data in Gb
Cum. Prod.: Cumulative production
Reserves: 2P reserves; data from Petroconsultants S.A
Y-t-F: Yet to find
Ultimate: Ultimately recoverable resource ('URR')
F95, Mean, F5: Probabilities
RG: Reserves growth in the 2P data
Source USGS (2000)

carry out various fossil fuel assessments, including Germany's BGR, France's IFP, the British Geological Survey and the World Energy Council, but the USGS assessments for oil and gas are by far the most extensive.)

In assessments carried out prior to the year 2000, the USGS had formally decided to exclude reserves growth in these calculations, recognising that while such growth appeared large in the US, this was fundamentally an issue with US data, and it would be inappropriate to apply reserves growth factors to 2P data for regions outside the US (see comments by C. Masters).

This changed with the Year-2000 Assessment, leading to the following data (Table A4.1):

As can be seen, a mean global allowance of 730 Gb was made for reserves growth, resulting in the total mean global URR for conventional oil of 3345 Gb, if both NGLs and reserves growth are included.

The USGS team in 2000 had accepted that their methodology for reserves growth was in part preliminary, based largely on US experience, so in work leading up to the 2012 Assessment a new approach was adopted. For the 2012 report they write: "*Unlike past assessments of reserve growth that relied on statistical extrapolations of growth trends, this [new] methodology includes detailed analysis of geology and engineering practices observed in developed fields*". But they add: "*Because of the paucity of data for many fields outside the United States, data acquired from U.S. fields undergoing reserve growth were used as analogs in this study*".

The resulting Year 2012 data are shown in Table A4.2.

As can be seen from Table A4.2, the Year-2012 Assessment puts global reserves growth of conventional oil (plus NGLs) as ~720 Gb; i.e., essentially unchanged from the Year-2000 value. By contrast, the Year-2012 Assessment puts the global conventional oil URR (including NGLs), if based on the approximations made here, at ~3850 Gb; i.e., about 500 Gb higher than the figure estimated in the Year-2000 Assessment.

Table A4.2 USGS Year-2012 Assessment: world conventional oil, including NGLs

Report date	Date of data			Y-t-F	Ultimate
2012	2009–11	Cum. Prod.	Reserves	Mean	Mean
Total ex-RG		*1250*[a]	*1110*[b]	*780*[c]	~*3150*
RG					723
Total + RG					~***3850***

Notes All data in Gb. RG: Reserves growth
– Though Year 2012 data are given by probability, this time these values have not been summed statistically as was the case in year 2000, so the 2012 totals are given only for the mean values (shown here)
– The USGS Year-2012 Assessment did *not* give data for global cumulative production or reserves, nor for yet-to-find in the US. Hence the following approximations have been made in compiling this table, with data not from the USGS year-2012 reports shown in the table in italics:
[a]Cum. Prod.: Based on the 2011 IHS Energy 'Liquids' value given in the text of Miller and Sorrell (2014), and as shown in Fig. 2.23, of 1248 Gb. Then a guessed 100 Gb subtracted to take off the non-conventional oil component in the IHS 'Liquids' category; and a further guessed 100 Gb added back on for production 2012–2015 inclusive
[b]Reserves: 2P reserves. Likewise, based on the 2011 IHS Energy 'Liquids' value given in the text of Miller and Sorrell (2014) for cumulative discovery of 2486 Gb, and with 1248 Gb subtracted off to give 2P reserves, as shown in Fig. 2.23. Then reduced to a guessed 1100 Gb to reflect only conventional oil
[c]Y-t-F: Yet to find. The USGS Year-2012 Assessment gives an estimate for the global mean undiscovered, *excluding U.S.*,. Including NGLs this is 732 Gb. Adding on a guessed 50 Gb for the U.S. (reduced from the specified 83 Gb for the U.S. Y-t-F in the Year-2000 Assessment) gives an approximate total for global yet-to-find, including the U.S. and NGLs, of ~780 Gb
– Note that all three of these estimates (Cum. Prod., Reserves, and Y-t-F), while certainly approximate, would probably agree with data the USGS would produce for these categories for conventional oil plus NGLs to perhaps +/− 50 Gb or so
USGS Sources
– *An Estimate of Undiscovered Conventional Oil and Gas Resources of the World, 2012*. USGS Fact Sheet: 2012-3028, March 2012
– *Assessment of Potential Additions to Conventional Oil and Gas Resources of the World (Outside the United States) from Reserve Growth, 2012*. USGS Fact Sheet: 2012-3052, April 2012
– *Assessment of Potential Additions to Conventional Oil and Gas Resources in Discovered Fields of the United States from Reserve Growth, 2012*. USGS Fact Sheet: 2012-3108, August 2012

Section A4.7, looks at the issue of how such a relatively high URR (at least, as compared to the URR data assumed by some other forecasters) should be treated.

A4.3.3 Reserves Growth: Summary

In summary, in terms of allowing for reserves growth in oil forecasting, we note that:

– Campbell has generally assumed reserves growth in the 2P data to be zero, maintaining that since the global peak is close, such gains—though real—will impact mainly after peak.

- By contrast, Laherrère does factor-in an amount for reserves growth.
- Harper of BP (& also Miller's early models) assumed a reserves growth factor of ~0.2 % annually globally might be reasonable, although this was before the current era of high oil price.
- Porter (1995) reported an annual increase also of about 0.2 % annually for US data from 1966 to 1979 (but where these data are apparently no longer being collected).
- Models that explicitly or implicitly use USGS URR data need to recognise the extent that reserves growth has been incorporated into these estimates.

As mentioned earlier, the only correct answer to the reserves growth questions listed at the beginning of this section will come from detailed studies by reservoir engineers in conjunction with other oil professionals. The topic is key because reserves growth is one of the main below-ground uncertainties (along with the true size of Middle East & FSU reserves, and the production rates expected of EOR techniques) of current oil forecasting.

A4.4 Contribution to Production from 'Fallow Fields'

Next we look at fallow fields. Globally there are many fields that have been discovered some time ago but not so far been developed, termed 'fallow fields'.

In some places, such as the UK, the fact that they have not been developed despite reasonable tax regimes, access to good technology, and a high oil price means they are mostly small or especially difficult in some way. For the UK, IHS Energy reports 800 field discoveries, twice the 400 reported by DECC; giving ~400 fields currently undeveloped, of which most in Laherrère's view will "likely never be developed".

By contrast, in some of the Middle East countries some of the fallow fields are large finds which are still unconnected to any pipeline, and which await political decisions on development.

Miller's results presented at an ASPO meeting in Vienna (Miller 2012) indicated that up to 170 Gb of global 2P reserves had not been developed by 2008. While clearly some of this was in appraisal, or awaiting sanction, Miller's tentative conclusion was that perhaps up to 140 Gb of the reserves are potentially doubtful, reflecting oil in the dataset which had not been developed for a very long time; oil which had seen production, but where this has not been recorded; or oil in fields now abandoned, possibly reflecting over-estimation of the original reserves.

All forecasters need to make judgements on when—and if at all—to bring on such fields. 'Capacity' forecasts, such as IHS CERA's, are perhaps more likely to assume such oil is available sooner, as does the 'bump' in the Miller forecasts; others may be more cautious on when they think such oil can or will be developed.

A4.5 Decline Rates and Depletion Rates

A *decline* rate measures the fall in oil production in a given year, for either a field or a region, expressed as a percentage of the previous year's production. Typically, it was the 'mainstream' forecasts that did not explicitly model the decline rate in existing fields that did not, as a result, forecast a conventional oil peak within their forecast horizons.

This changed once decline rates were examined quantitatively, and some good studies on decline rates have now been carried out for individual fields and for regions; and also for fields post-peak, and for all fields (where in the latter, some fields may be at plateau and new ones ramping up). These studies include those from the IEA, Uppsala University and IHS Energy; for details see the UKERC report (Sorrell et al. 2009). It goes without saying that forecast models need to use values for field or regional decline that are realistic.

The *depletion* rate is not so unambiguously defined, but here means the fall in production in a given year expressed as a percentage of the remaining recoverable resource, sometimes abbreviated as 'DRRR'. If decline is exponential, then the decline rate and the depletion rate are the same.

Aleklett (2012) took issue with an apparently high depletion rate assumed in the IEA's 2008 *World Energy Outlook* for fields yet-to-find (hence allowing fast production). There may be no intrinsic upper limits to DRRR, but Aleklett and colleagues are right to point out that depletion rates that are out of line with past experience need justification.

A4.6 Applying a 'Hubbert Curve' to a Region or the World

Applying a Hubbert curve to a region can mean a number of rather different things:

- It can mean fitting a hand-drawn curve to past production in a region and at the same time projecting production forward to meet a given estimate for URR, as Hubbert did in his initial forecasts.
- Or it can mean matching one or more (symmetric) logistic curves to the region's past production, again meeting an estimate (or estimates) for URR; and where this fitting can be by eye, or done statistically.
- Or by again matching one or more logistic curves to past production, but now letting the projection of the curve over time itself give the estimate for URR; i.e., requiring no exogenous estimate of URR.

The first two methods work well for regions where production has been relatively unconstrained, and where the assumed URR is not too optimistic (based, for example, on extrapolated discovery).

The third method can in theory be applied almost from the earliest date of production, and works well provided production does indeed follow a logistic curve. But in practice (see the examples in Campbell's *Atlas*, 2013) it tends to work well only towards (or after) the production peak, i.e., once a logistic shape has settled in. This third method can be done by eye, or statistically, and in the latter case the linearisation of the Hubbert curve to project the URR by a straight line can be very effective. For examples, again see Campbell's *Atlas*; also the section on Deffeyes in Annex 5; and the detailed plots in Hallock et al. (2014).

Note also that such a 'production-only' method (i.e., not using an exogenous value for URR, either for the region under consideration or for specific fields within it) has to be handled carefully if applied to regions where production has been significantly constrained, perhaps for political reasons; and including for the World as a whole, where here production was constrained by reduced demand, triggered by the high oil price shocks of the 1970s.

In the latter case, as Fig. A4.2 (left chart) shows, the fit between a simple Hubbert curve (here for a URR of 2100 Gb) and actual production becomes very poor following the first oil shock in 1973. Figure A4.2 (right) shows one global forecast (that of Campbell) to indicate how the relatively sharp peak of the Hubbert

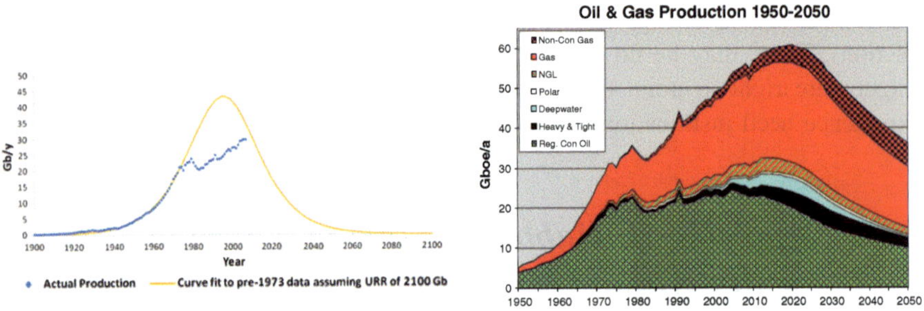

Figure A4.2 Global oil production. Showing (*left chart*) the very poor fit between a single logistic ('Hubbert') curve and actual global oil production; and (*right chart*) comparison to possible future oil production (Campbell forecast).

Left chart Global oil production 1900–2100: Actual production: dots. Fitted logistic curve with URR of 2100 Gb, line.

Source Probably UKERC (2009) report.

Right chart Forecast by Campbell for global oil and gas production: Actuals: 1950–2012; Forecast 2013–2050. Shows 'All-oil' (excludes GTLs, CTLs and biofuels) and 'All-gas' (excludes UGC and methane hydrates). Legend (reading from top of graph): [Gas]: Non-conventional gas; Gas; NGLs. [Oil]: Polar; Deepwater; Heavy and Tight; Regular conventional oil. ('Blip' at 1991 is losses from the Kuwait oil fires).

Source Campbell, March 2014.

The oil shocks of the 1970s increased the price of oil, and hence reduced its production growth dramatically. It is possible that had the 1970s price shocks not happened, and the pre-1973 growth pattern continued, then global oil production would have followed the Hubbert curve at least reasonably well, and global oil production of conventional oil be now be about two decades post peak curve in Fig. A4.2 (left) in fact spread out substantially, reflecting the impact of the reduced supply of cheap oil post the 1970s oil shocks, and consequent higher prices. Even so, as Fig. A4.2 (right) shows, overall, global production, while still not too far from a 'peak at mid-point' shape, is now closer to the 'undulating plateau' anticipated by IHS CERA.

A4.7 Use of URR Estimates

As indicated in the previous Section, some oil forecast models use URR estimates explicitly, while others effectively generate a URR by forecasting future production. (The latter, when added to cumulative past production, plus some amount for assumed yet-to-find at forecast-end and anticipated future recovery gain, yields an implicit URR.)

Where a URR estimate is based on extrapolated discovery (for example, from plots of cumulative 2P discovery against date, or better, against new field wildcats —'creaming curves'—as used by Campbell, Laherrère and others), then for conventional oil a regional peak at ~50 % turns out to be a reasonable assumption (though in practice most regions now past their peak of conventional oil have peaked at somewhat less than this).

If instead a 'global assessed' value for URR is used that includes assumed reserves growth, as in the USGS year 2000 and year 2012 estimates for example, then these URR estimates are typically significantly higher than 'extrapolated discovery' estimates, and the peaks of conventional oil production in regions past peak have typically occurred in practice at percentages quite a bit lower, nearer 25 % or 30 %, of such assessed regional URR values.

This is borne out by the right-hand plots of Figs. 2.6, 2.9, 2.12 and 2.15, which illustrate how USGS year-2000 URR estimates without reserves growth look compatible with backdated 2P discovery data, while if reserves growth is included (e.g., Fig. 2.21 right, for the World) then the assessed global URR looks high in terms of any likely discovery pattern.

As mentioned in the discussion above on reserves growth, USGS assessments earlier than year 2000 explicitly discounted significant reserves growth in regions outside the US. But the year-2000 estimates applied a US-based formula to estimate reserves growth across the other regions of the globe. Despite the USGS providing a clear caveat as to how these data were generated, the data became 'carved in stone' and went on to be used widely, for example in a number of IEA *WEO* forecasts which quoted the USGS year-2000 'grown' mean estimate for the global conventional oil URR (including NGLs) of 3345 Gb.

As Table A4.2 shows, the corresponding year-2012 estimate is probably in the region of 3850 Gb for conventional oil plus NGLs; while the EIA data in Table 3.1 gives remaining global conventional oil reserves as 3012 Gb; which, with ~1250 Gb added for conventional oil produced to-date, gives an apparent global conventional oil URR (probably including an allowance for reserves growth) of ~4250 Gb.

It can be argued that such 'relatively high' URR estimates do indeed represent realistic possible *long-term* values for the global 'ultimately recoverable'; it is hard to know what global average oil recovery factor might eventually be achieved. But, as mentioned above, URR values that are significantly out of line with the current 2P discovery trend must be applied with caution if correct near- and medium-term forecasts are to result.

It thus seems that the primary cause of differences between the current various oil forecasts lies in the different values they assume (or generate) for the global URR for conventional oil. This is therefore a critical technical issue on which oil forecasters should aim for consensus; for greater detail see Bentley (2015b).

A4.8 Field-Aggregate Forecasting Models that Do Not Need a URR Estimate

This section summarises briefly some field-aggregate oil forecasting approaches that do not need an external estimate for a region's URR, and thus which avoid the controversy over what is the region's 'real' URR.

One such approach, as mentioned above, is 'Hubbert linearization'. Here, production divided by cumulative production is plotted against cumulative production, with extrapolation of the resulting line (typically roughly straight, later in a region's development) giving an estimate for URR; and where date of peak can be estimated from the mid-point of this URR. This technique was used by Hubbert, and later by Deffeyes, Campbell, Laherrère and Hallock et al., among others.

Another approach is PFC Energy's simple rule-of-thumb that production in a region typically peaks when cumulative production in the region reaches about 60 % of cumulative 2P discovery in the region (PFC Energy 2005). This is a fairly robust rule, and is easy to apply.

A third such approach is that of plotting production as a 'mirror' of the discovery curve, where the latter is delayed by a number of years that is dependent on the region under consideration. This approach works acceptably for both 1P 'discovery' data, as Hubbert showed for the US, where the cumulative production curve fairly accurately mirrors 1P cumulative 'discovery' curve lagged by 10 years (Fig. A2.2); and with 2P discovery data, as applied by Laherrère to many regions, even though 2P discovery curves often rise quite steeply early on, as the big fields are found.

A4.9 Bottom-Up By-Field Models: Capacity Versus 'Realistic'

In examining oil forecasts, it is important to distinguish 'capacity' from 'realistic' oil forecasts models. Some bottom-up models forecast how much oil could be produced under ideal conditions, while others seek to reflect a range of above-ground factors to generate a more cautious forecast.

IHS CERA, for example, says its model (Fig. 4.1) aims to capture much of the above-ground reality (timing of investment needed to bring projects forward, slippage of projects, availability of drilling rigs and so forth) but is explicit in stating that its forecast is of 'capacity', that is, forecasts what could be produced if all technically available oil is produced, subject to the above provisos.

This however ignores what everyone agrees are difficult-to-model above-ground factors, such as the willingness of oil exporters to produce now rather than in the future. It is also possible that IHS CERA's forecast may be overly positive on the size of Middle East reserves, and on the world's ability to develop its fallow fields.

By contrast, Globalshift Ltd.'s model (Fig. 4.3, left) aims to incorporate these factors, and hence produce a more 'realistic' forecast.

Annex 5
Additional Details on Specific Oil Forecasts from Individuals and Organisations

This Annex gives additional details on approaches to modelling oil supply by specific individuals and institutions, supplementing the information in Chaps. 3 and 4.

Details are given for Hubbert, Ivanhoe, Campbell, Laherrère, Petroconsultants, Deffeyes, Smith, Skrebowski, BGR, LBST, Uppsala University, UKERC, IEA, EIA, oil companies (Shell, ExxonMobil, Chevron, ENI, Total and BP), and banks and investment houses. These individuals and organisations are discussed because of the light their forecasts shed on the 'peak oil debate'. Others could have been included, as quite a few organisations carry out oil forecasting.

Note that some of the information provided below is based on personal experience and on conversations with others. There is always a degree of uncertainty associated with such information, so some statements in this Annex should be treated with a degree of caution. Where readers can offer improved information or insight into how the peak oil debate evolved, this would be welcome.

A5.1 Hubbert

Hubbert's work has been discussed in Chap. 4, but here is amplified, in particular on the evolution of his forecasts for global oil production, and in giving some of the methods he employed to estimate URR. Also included are comments Hubbert made in an interview with Steve Andrews which shed important light on some of the details of Hubbert's thinking.

(i) **Evolution of forecasts, 1956–1980; and methods to estimate URR.**

<u>1956: Peak derived from an industry estimate of 'ultimate' plus a hand-drawn production curve.</u>

(a) US Lower-48 oil production.

In 1956 Hubbert obtained estimates from industry sources for the original endowment (URR) of recoverable conventional oil of the US Lower-48 states, including continental shelves. These estimates were 150 and 200 billion barrels

R.W. Bentley, *Introduction to Peak Oil*, Lecture Notes in Energy 34,
DOI 10.1007/978-3-319-26372-4

(Gb). He checked these within Shell where he worked, and then drew two production curves by hand that matched US historical production up to 1955 and which encompassed total production volumes (past and future) given by the estimates for ultimate. The curves showed that US production would peak either about 1965 or 1970, depending on which ultimate was used (Hubbert 1956). As is well known, US Lower-48 production peaked about 1970.

As Hubbert wrote later (Hubbert 1982, p 41, paraphrasing slightly), '*The predicted production curves used in 1956 were not based on an empirical equation or assumption of symmetry, but in accordance with the constraints on matching ultimate and the necessity that the decline be gradual and asymptotic to zero.*' Moreover he noted: "*That the production of exhaustible resources does behave in this way can be seen by examining the production curves of some of the older producing areas*" (see e.g., Figs. 12 and 13 for Illinois and Ohio in his 1956 paper).[15]

Hubbert also pointed out that under this 'gradual decline' assumption there was almost no choice as to how the 150 Gb ultimate production curve could be drawn, as by 1955 52 Gb had been produced, and there were 30 Gb in known proven reserves. Even the 200 Gb curve had little scope for a different profile if it was to look like the profiles of regions already past peak.

Although this approach is essentially empirical curve drawing once an ultimate has been estimated, the importance of Hubbert's geological understanding in advocating this 'peak and decline' model should not be overlooked. He understood that loss of reservoir pressure causes the decline in individual fields; he knew that large regions tended to have a single peak; and he pointed out the geological realities that can make for complex production profiles in specific cases.[15] As mentioned in Chap. 4, the significance of the 1956 forecast was that most observers at that time assumed that any potential US shortfall would be far into the future, presumably based on the on-going exploration successes and the long US history of proven reserves replacement. Even those more familiar with the data were perhaps misled by the large 1956 US 'yet-to-produce', of between ~100 Gb and 150 Gb, versus the cumulative produced of only 52 Gb.

Today the estimate for the conventional oil ultimate of the US Lower-48 (ex-NGLs), if based on discovery trend, is about 190 Gb. The 1956 estimates given to Hubbert were therefore pretty accurate. This was simply because the 'P50' (i.e. 'reserves-grown') discovery rate of the Lower-48 states had peaked back in the 1930s, as is clearly seen in Fig. 15 of Hubbert (1967), so by 1956 nearly all the Lower-48 oil that could affect the production peak had been found.

In summary, Hubbert's 1956 approach used geological knowledge, experience from regions already past peak, and estimates of ultimate from recognised sources to show that the US peak was predictable, and fairly close.

(b) World oil production

In this 1956 paper Hubbert applied the same hand-drawn 'peak and decline' technique to the world. For the estimate of global conventional oil ultimate (excluding NGLs) he used mainly data from ESSO's L.G. Weeks, but increased this based on new data from the Middle East plus USGS information, to give an ultimate of 1250 Gb. To fit this ultimate he drew an asymmetric global production curve that tailed away to the right, and which peaked at an assumed 2.5 times then-current production, so peaking at 35 Mb/d. This curve placed the date of global peak "at about the year 2000".

This plot nicely demonstrates a strength of this 'Early Hubbert' approach. Today the global conventional liquids ultimate (i.e., including NGLs) as it affects peak is estimated to be in the range 2000–2500 Gb depending on assumptions, giving a peak between about the year 2000 and 2010. This ultimate is nearly twice the oil-only figure used by Hubbert in 1956, but the date of peak is little different. With this technique the date of peak is only little sensitive to the size of ultimate, provided the latter is not wildly wrong. On the other hand the method requires an estimate for ultimate, and peak's *height* (as opposed to its date) is fairly sensitive to the assumed ultimate.

In explaining the 1956 estimate of world ultimate it is worth pointing out that the full size of Middle East reserves was only then starting to become known (in Hubbert's words, to a "*rough order-of-magnitude*"). More importantly, the world discovery peak was still a decade into the future (in the mid-60s). As pointed out earlier, it is only when discovery declines that one be more sure about how much the drill bit is likely to find. Finally, the fact that in the US (and some other places) published oil reserves are only very conservative 'proven' data means that the recoverable resource base will be underestimated unless quite large 'reserves growth' factors are applied.

Three additional points from this 1956 paper should be made. On US conventional gas, Hubbert's 1956 prediction was significantly conservative in volume, although not far wrong on date of peak; he used the estimate for the US gas ultimate of 850 Tcf from Pratt. A second, crucial, point is that Hubbert knew that at that date the recovery factor was only about one-third of oil in-place. He recognised that recovery would improve, but stated: "*... it appears unlikely that any improvement that can be made within the next 10 or 15 years can have any significant effect on the date of [peak]. A more probable effect of improved recovery will be to reduce the rate of decline ...* ". The situation is the same with the world peak today. Thirdly, Hubbert quoted the estimates by others for the amount oil likely economically extractable from world deposits of tar sands and kerogen shales; these still look about right at 800 Gb and 1000–2500 Gb respectively.

1962: Peak derived from historical data on proved discoveries, proved reserves, and production; use of the logistic curve to estimate ultimate. Also ultimate derived from field distribution by size and number; large fields 'grown' and backdated.

(a) US Lower-48 oil production.

Subsequent to Hubbert's 1956 paper, estimates for the US Lower-48 oil ultimate grew significantly, up to as much as 600 Gb—pushing the anticipated date for the US peak in this case out to about the year 2000. Some of these estimates were based on intrinsic basin 'oiliness', the so-called 'geologic' method.

In response to the uncertainty introduced by high estimates for the US ultimate, Hubbert devised a way of using only public-domain historical oil information to determine the date of peak (Hubbert 1962). He plotted US cumulative production, *proved* reserves, and cumulative 'proved discovery' against time, and examined these curves as well as their derivatives. (As Hubbert recognised, 'proved discovery' is a derived variable, calculated as cumulative production plus proved reserves.)

The general pattern to be expected for a region is that discovery peaks, then reserves, then production. (But if only conservative proved reserves are available one has to be cautious about depending on this sequence.) In the US in 1962 the curves contained enough data to show that 'proved discovery' had indeed peaked, and that proved reserves were just about at peak. Plots of cumulative proved discovery and cumulative production both showed near-logistic shapes, with the latter curve 'mirroring' the former lagged by about 10 years. Taken together the data showed that peak production must be close at hand, with the derivatives of these curves supporting this analysis. Hubbert noted that both the ultimate and date peak could be determined with this method once "*production in a region has gone through about the first third of its cycle*".

To be more precise about peak, Hubbert tried fitting a variety of curves to the production and 'proved discovery' historical data. He found that most curves were unusable, but perhaps from a technique used elsewhere in the mineral industry, discovered that the logistic curve answered very well. This allowed him to treat the curves analytically, where he developed an iterative technique to determine ultimate. This technique plotted, against time, historical values of a log function that divided a guessed ultimate by cumulative production, and looked for a linear fit. The latter gave the US Lower-48 ultimate as 170 Gb.

The key advantage of this 1962 approach was that no estimate for ultimate was needed *a priori*; both the ultimate and the date of peak fell out of simple production and reserves data.

In this report for the National Academy of Sciences, Hubbert used a second method to support the above analysis. This examined the quantity of oil discovered in large fields, as reported in the *Oil and Gas Journal*, plus that in small fields. For these calculations, Hubbert recognised that a field's proved reserves grew over time, from a very conservative initial estimate (reported under SEC or earlier rules), by the addition of accumulated 'extensions and revisions'. For the large fields, which contain the bulk of the oil, Hubbert applied a time-lapse formula to 'grow' the fields (i.e., adjusting for expected future growth in individual field sizes) and backdated these data by plotting field final volume—in this case membership of the 'large' category—against date of original discovery. Summing long-term expected oil in both large and small fields gave the estimate for the Lower-48 ultimate as 175 Gb. This provided strong support for the 170 Gb figure derived from the historical production data.

Note that these estimates of ultimate in 1962 were generated when the US had produced 67 Gb, and its proved reserves stood at 32 Gb.[16]

(b) World oil production

For this 1962 report, for calculating the date of the world peak Hubbert stayed with the estimate for world ultimate of 1250 Gb mentioned above. He drew an asymmetric graph as in 1956, and again predicted peak at 35 Mb/d "*about the year 2000*".

1967: Examination of grown backdated discovery data. Also estimating ultimate from grown backdated discovery per foot drilled.

In a later paper (Hubbert 1967), Hubbert represented the analysis of US historical data on proved discoveries, proved reserves and production given in the 1962 paper. He then used newly available field reserves growth data to examine the discovery history more rigorously in 'grown' backdated form. This allowed the realistic discovery trend to be clearly seen. Whereas 'proved discovery' (cumulative production plus the conservative proved reserves) for the US had peaked in the mid-1950s, backdated grown discoveries ('true'; or idealised 'P50' discoveries) had peaked in the mid-1930s. Hubbert did not extrapolate this grown backdated discovery trend to get an ultimate, but showed that its fall-off against time made an ultimate in excess of 200 Gb unlikely.

He then examined a powerful new method for estimating ultimate that had been developed recently by Zapp (1962). This extrapolated discovery data per cumulative exploratory footage drilled. Its advantages were a greater independence from economic factors (removing time and oil price from the discovery trend); and the fact that it could set an 'end to exploration' by physical limits (an assumed one exploratory well per 2 square miles, and a drilled depth of either 20,000 ft or the basement), rather than by the unknown date of the very last well. However, a problem with this method was that it seemed to generate much larger estimates for the US ultimate, of up to 590 Gb for the Lower-48 if marginal and sub-marginal resources were included; and a US ultimate of 660 Gb if Alaska was included.

After some investigation Hubbert found the error: that these estimates implicitly assumed the volume of oil discovered per exploratory foot would stay approximately the same into the future. Hubbert saw that this could not be the case: in future, smaller fields and deeper fields were both likely to need many more exploratory feet to discover than large or shallow fields. He analysed this aspect in detail, developing a new formula to 'grow' the size of the fields discovered, and backdating these data to the original date of discovery. This showed that discovery per exploratory foot had indeed declined drastically from its peak in the 1930s (Fig. 15 of Hubbert 1967), with extrapolation giving the Lower-48 ultimate as between 150 and 170 Gb.

Hubbert noted that the technique implicitly incorporated future improvements in technology, only constraining these to occur no faster than had been the case in the past. He concluded that "*ultimate US crude oil production* [Lower-48, ex NGLs] *as large as 200 Gb would be difficult to justify*".

1969: 'Energy Resources' paper—World oil production

In 1969 Hubbert used two estimates for the world ultimate, of 1350, and 2100 Gb. In each case he drew a symmetric derivative-logistic shaped curve (the 'Hubbert' curve), and predicted the world peak at 65 M/d in about 1990 for the lower ultimate; and at about 100 Mb/d in about the year 2000 for the higher ultimate (Deffeyes 2005).

1977: Using Nehring's estimate of 2000 Gb for the world ultimate.

Around 1977, a new estimate of the world's conventional oil ultimate became available. As *The Global 2000 Report to the President* (Barney 1982, p352) notes:

> *A recent report prepared for the CIA by Richard Nehring of the Rand Corporation provides a detailed, publicly available description of the known recoverable crude oil resources of the world and an explicitly reasoned estimated range of ultimately recoverable conventional oil crude oil resources.* [This] *focuses on the relatively small number of giant oil fields [>500 Mb, which] contain more than 75 % of the known recoverable oil resources of the world [and] concludes that the ultimate recoverable conventional crude oil resources of the world are somewhere between 1700 and 2300 billion barrels.*

Hubbert took Richard Nehring's "*best estimate*" for the ultimate from his range, of 2000 Gb, and matched global historical production up to 1974 to two possible production curves representing this volume. One was an unconstrained derivative logistic curve, which peaked at about 100 Mb/d in 1996. The second curve examined how long the recoverable resource base would last if production were held at the 1974 level. This showed that production could stay flat until about 2035, after which it would decline quite steeply, reaching near-zero by 2080 (Barney 1982, p 353. This references: 'M. King Hubbert in Congressional Research Service, Project Interdependence: *U.S. and World Energy Outlook Through 1990*; Washington, 1977, p 642.').

In the event, demand followed neither of these postulated curves. The high oil prices of the late 1970s forced demand significantly below the unconstrained logistic curve and so pushed back the date of peak. If Nehring's best estimate for ultimate is correct, today the peak lies at about 2010. Note that when Hubbert produced these projections, only 340 Gb out of the assumed 2000 Gb had been produced. Today ~1250 Gb has been produced.

1980: Proceedings of a Symposium at the Department of Commerce, Washington DC, June 18–20 1980 (published 1982).

This is Hubbert's best overall summary paper, and covers a great deal of ground (Hubbert 1982).

He first examines the 'geologic' (intrinsic basin rock 'oiliness') method for estimating ultimate, and shows that this is of little use, as oil richness can vary by a factor of 100 between basins, and there is no way of knowing this factor before exploration is significantly advanced.

He then looks at the typical complete cycle of oil production in a region, and emphasises (p 34) that: "*There is no necessity that* [production has] *a single maximum, or that it be symmetrical. In fact the smaller the region, the more irregular the shape of the curve ... For large areas, such as the entire United States or the world ... the irregularities of small areas tend to cancel ... and the curve becomes a smooth curve with only a single principle maximum.*" But he notes that even such a 'large-region' curve need not be symmetric.

He then examines the mathematics to derive the production cycle curve from primary data. This work probably follows on from his earlier discovery that the logistic curve was a good fit to US data, but here he derives the logistic curve from a simple assumption on the shape of production versus cumulative production. This leads directly to the iterative linearisation described earlier, and with integration gives the conventional logistic formula. He also puts this in the form of a simplified linear relationship by plotting production divided by cumulative production against cumulative production. This turns the logistic curve into a straight line that intersects the abscissa at the value for ultimate.

He then re-presents the 1962 method for determining peak that uses simply historical production, proved reserves and proved discoveries data. The additional data available by 1982 had extended the curves considerably from the original 1962 analysis, and with the US peak by now well past, confirmed the correctness of that analysis. It is crucial to note that in using proved reserves Hubbert (p 59) stressed that these represent "*oil in existing fields that has been proved by development drilling and is recoverable by existing installed equipment and technology.* [It] *is a conservative figure and is not intended to represent the ultimate amount of oil that known fields will produce.*"

He then also reprises the third method for determining ultimate and peak, based on discoveries per cumulative feet drilled. Here he re-examines the methods by which future increases in field size can be estimated, so that 'grown' (and back-dated) data can be used in this 'discovery-per-cum.-ft.' method. Again, the updated data available in 1980 (from D.H. Root; see Fig. 38 of Hubbert 1982) confirmed the dramatic fall-off in US 'true' oil discovery since the 1930s that had been reported in 1967.

The paper then used similar methods to analyse future US gas production, and finishes with a report of the meeting's follow-up discussion.

The latter is particularly revealing. Hubbert again stresses that the production curve need not be symmetric; he states explicitly that his study is of conventional oil, excluding "oil shale [kerogen oil], coal, and the Orinoco heavy oil"; he recognises the scope for increased recovery but notes that "*the effect may be easily exaggerated*", and in terms of increased recovery he brings in the notion that energy content versus energy for extraction sets a physical limit on what will be extracted for energy use. Finally, and perhaps most importantly, he addresses the issue of the effect of oil price in increasing oil supply. He says: "*During the last decade we have had very large increases in the monetary price of oil. This has stimulated an accelerated program of exploratory drilling and a slightly increased rate of discovery, but the discoveries per foot of exploratory drilling have continuously declined from an initial rate of about 200 barrels per foot to a present rate of only 8 barrels per foot.*"

How should we summarise Hubbert's work? In the abstract for this 1980 paper Hubbert says: "*Methods based on this type of analysis, which have been developed and used by the present author during the last 25 years, have consistently given predictions of the future course of oil and gas production in the US which have agreed within narrow limits with what has subsequently occurred*". There is no question that Hubbert, with data and help from others, introduced powerful new methods for predicting future hydrocarbon production in a region. That he did so in a country where reserves data are so misleading, imminence of the production peak little recognised, and opposition to a prediction of peak often quite strong, is a major testament to the man's ability.

(ii) **Hubbert interviews with Steve Andrews** (see also Campbell, Ed. 2011, Chap. 2).

Hubbert was interviewed in March 1988 by Steve Andrews, later one of the founders of ASPO-USA. Some of the many informative remarks in this interview include (Hubbert speaking):

> ... *We know that [the oil production] curve is going to go over the hump and go back to zero. After the hump, it can't drop sharply; it's got to go down gradually because that's how you find oil. You don't find it all at once and you don't produce it all at once. It peters out gradually.*

> ... *Between 1966 and 1971, I was using commonly accepted oil data, the best data available.*

> ... *In a typical bull session in the industry, the question would often come up: how long will it be before we hit the peak of our [US] production? The intuitive judgment: here we've been in the oil business 100 years and produced a whisker over 50 billion barrels. For the estimates between 150 and 200 billion barrels, the intuitive answer was 'not in our life-times.' Our grandchildren may have to worry about it.*

> ... *After my talk* [on the nearness of the US peak], *there were two camps of thought. The first reaction was 'the guy must be crazy.' It violated their intuitive judgment. Everyone went back to take their best look at this quantity* [US Lower-48 conventional oil URR]. *Some found it fell within their range of estimates and they couldn't avoid the conclusion. Others found the conclusion so abhorrent that they couldn't accept it.*

> ... *You don't change the implications of that 1970 peak production curve by small figures; it takes big ones. Within a year the figures* [for US conventional oil URR] *began to escalate. At the end of five years, they had more than tripled.* [The latter refers to estimates, perhaps *inter alia*, by Zapp, mentioned above, for the USGS of up to 590 Gb for US Lower-48 oil (Zapp 1962), and later for all-US oil of up to 587–620 Gb (Moore 1971).]

(iii) **Additional information re Hubbert's oil forecasting**:

In addition to Hubbert's papers and other references, further information on Hubbert's forecasting can be found from:

- Presentations by Hubbert on YouTube.
- The Steve Andrews' interview mentioned above.
- Ron Swanson's excellent website: www.hubbertpeak.com.
- A UK Open University class video, with introduction by L. Ivanhoe. Title: *Oil: Finds for the Future*, produced for the course 'The Earth's Physical Resources', S238; 1984.
- *The Oracle of Oil*; forthcoming book by Mason Inman to be published by W.W. Norton.

A5.2 L.F. Ivanhoe

L.F. 'Buzz' Ivanhoe was an experienced petroleum geologist who early on had argued against the analysis of B. Grossling of the USGS, where the latter suggested that the quantity of undiscovered oil was large because many regions in the world had not been drilled to the same extent as the US. He gave a knowledgeable commentary in the UK Open University film mentioned above on peak oil that included a contribution from Hubbert, and he and Hubbert were in close communication for the last five years of the latter's life.

Also it was Ivanhoe who warned Campbell of the 'quota wars' step-changes in OPEC 1P reserves. Ivanhoe sponsored and published a quarterly newsletter for journalists and policymakers on the topic of peak oil through the Colorado School of Mines from 1996 to 2002. He wrote many articles for the *Oil & Gas Journal* and *World Oil*, including an excellent paper on future global oil supply based on production mirroring discovery (Ivanhoe 1996).

A5.3 Campbell

Campbell's work has been discussed in Chap. 3. See also Campbell's many publications, and also Hall and Ramírez-Pascualli (2013).

Campbell is always clear to say that the data available to make oil forecasts are unfortunately poor, and that the key point is not the exact date of peak, but the economic and societal stresses that result from subsequent decline.[7]

For 'Regular conventional' oil (see below, and Chap. 3) Campbell models—in general—the combined production of all fields in a region (usually a country) as follows:

- If a region's production is past midpoint he assumes that future production declines at the current depletion rate (defined as annual production as a percentage of his assessed total future 'Regular conventional' oil to be extracted in that region by 2100). In this model, for 'regular oil' most countries save in the Middle East Gulf are past their midpoint and into decline.

- For countries prior to mid-point, production grows at the current rate until the region's mid-point of URR is reached, after which production is declined away exponentially at the depletion rate valid at the mid-point for that region.
- For the Middle East Gulf countries, he assumes a plateau to midpoint and then decline at the then depletion rate.

'Regular conventional' oil is defined as crude oil, but taking out the following categories of oil for the reasons stated:

- Deepwater (>500 m water depth): This oil is seen by Campbell as intrinsically expensive to access because of high energy requirements, and as a result is generally produced rapidly, and only from large fields.
- Polar: Will only come on slowly because of the constraints of working in the Arctic. (Campbell notes: "… *also the geology* [of the Arctic] *is not very favourable due to large vertical movements of the crust under the weight of fluctuating ice caps. They tended to depress the source rocks into the gas window. So the flukes necessary for oilfields such as Prudhoe Bay are rare.*"
- Heavy and tight oil: Tar sands is '*more like mining*'—investment led, but no *intrinsic* limit on rate other than overall size of recoverable resource, and energy limits as poorer sources are accessed; Orinoco oil has its own constraints of cost and politics; light-tight has constraints of access outside the US, and unusual production profiles: very fast decline and long tails at each fracking and re-fracking.
- NGLs: Tied to each region's forecast gas production profile.

Campbell comments:

– I no longer refer to "Reserves" as such, but instead to Future Production (to 2100) with a certain percentage being "Known". This basically refers to so-called 2P Reserves. The balance is classed as Yet-to-Find, which could include so called Reserve Growth. The engineers rightly took a very conservative view of the size of a discovery because they faced massive investments to develop it. Naturally over time their estimates were raised, and they got medals for their technological skills. Us explorers on the other side of the coin often had to exaggerate the size of prospect to get approval to drill it under a limited corporate budget, facing many other claims.

– It is a bit hard to know how to deal with fallow fields. Companies do not drill a prospect unless in the success case it would be profitable, but as I mentioned above they might have been misled by the explorer's exaggeration. If a distant non-economic discovery is made somewhere it is rightly not included in the reserves of the country. But naturally some of them might later become viable with higher oil prices and the construction of pipelines to other fields in the vicinity. I think whatever the fallow fields may ultimately deliver might as well be treated as Yet-to-Find.

[For completeness, here are some comments from Campbell that give a further insight into his views:

On proved ('1P') reserves, Campbell writes: "*I think 1P reserves are based on what current wells are expected to produce, ignoring infill wells and fringe wells drilled late in a field's life. Another problem is to define an oilfield. A single field can have different names if it crosses a lease boundary or frontier (the North Field of Qatar becomes South Pars in Iran), furthermore many fields have subsidiary traps found late in life. Many US New Field Wildcats should be classed as New Pool Wildcats. It touches on the issue of the unique combination of geological circumstances that allowed oil to accumulate in a particular place, albeit in several different structural traps and reservoirs. ... To a large degree [proved] reserves was a financial term, only partly related to what was actually recoverable in Nature.*"

On use of reserves data, Campbell notes that: "*The best estimate for the industry was normally proved reserves in full,* [plus] *two-thirds of the probable and one-third of the possible.*"

On oil price, Campbell writes: "[This] *is a critical issue, because high prices prompt recession, reducing demand—and* [lead to] *all manner of political tensions, as in the Arab Spring, Iraq, etc. So I don't think that rising shortages will lead to prices above say $150* [per bbl] *in to-day's money. Rampant inflation is a likely outcome, as the most practical way for governments to remove all the money they printed out of thin air on false assumptions of eternal growth.*"

On the significance of the overall topic, Campbell concludes: " *... I think that there might be an additional aspect to cover. It is that throughout history, it is energy supply that controlled empires. In early years the energy came from food. The Romans grew it and expanded, feeding their soldiers, but gradually their own fields were not enough, perhaps due to soil erosion and cutting down forests, and they had to invade France for more land. But eventually it peaked and went into decline. It was the same story with other empires such as the Mayans or the Anasazi people in Colorado. Food storage between harvests provided a critical control, laying the foundations for banking. This is all brilliantly described ... in a book* Energy Civilization—The Zenith of Man *by Douglas Reynolds. ... That is why Peak Oil matters so much, spelling the 'zenith of mankind' in his present status. It speaks of a compelling need to reduce population, and already we see growing tensions around the world as people blame their governments for the harsher conditions with rising food costs and unemployment. That basically is at the heart of the Peak Oil story.*"]

For a recent description of Campbell's oil and gas forecast model, which now is perhaps the first to include the important aspect of net-energy in terms of oil and gas production, see Campbell (2015).

A5.4 Laherrère

In a somewhat different approach, production can be modelled by summing a sequence of logistics curves, where each curve represents a different 'source' of oil, such as conventional oil from a later exploration phase in a basin, and also—as Laherrère uses in his global model—different categories of oil such as tar sands and other non-conventionals.

Laherrère has single-handedly been the source of many of the most valuable graphs in the entire peak oil debate; see his many publications (often online), the graphs here, and in Hall and Ramírez-Pascualli (2013). Note that in looking at these graphs, one needs to bear in mind (as in the case of the Campbell data) the categories of oil included, and any adjustments Laherrère has made to industry 2P data, for example in reducing Middle East reserves and FSU data.

As Laherrère's papers show, he has worked on a wide variety of techniques for oil forecasting, including introducing the parabolic fractal for oil field size distribution, mirroring production to discovery with time-lag, and adding multiple logistic-derivative (Hubbert) curves, as mentioned above.

As with Campbell, Laherrère's predictions are not purely data based, but rely also on an immense amount of knowledge and experience of oil geology. For a recent description of Laherrère's 'all-oil' forecast model, including background on how he came to see the need for such modelling while working for Total, see Laherrère (2015).

A5.5 Petroconsultants

The Petroconsultants 1995 report *The World's Supply of Oil, 1930–2050* generated assessments of individual oil ultimates by country, and calculated the consequent rates of future production, and summed these for global production. The report was based on a 'petroleum system' approach (source rock, burial history, migration, reservoir and seal), and URR estimates for conventional oil (ex-NGLs) were made for a wide range of countries on the basis of a number of techniques.

The general methodology was as follows:

- Estimation of 'P50' oil reserves by country. ('P50' reserve estimates are those with a notional 50 % probability of being correct, and approximate to 2P estimates). These estimates were generated by taking ultimately recoverable *field* reserves ('URR') data from the Petroconsultants' database, but adjusting these in the light of the authors' geological knowledge, and on the basis of reasonableness tests. A key test was to plot field production versus cumulative production. This linearises exponential decline, and for most fields where decline is underway provides a good indication of the field's likely URR. This approach showed, for example, that many field reserves in the former Soviet Union appeared to be over-reported in the dataset.
- Generation of estimates of oil yet-to-find. This analysis mostly used a range of statistical approaches, largely based on discovery data to-date, to estimate the quantities of conventional oil likely be found in the region within a reasonable exploration time-frame (for example, from twice as many wildcats as already drilled). The techniques included plotting regional 'creaming curves' (plots of cumulative discoveries versus new-field wildcats); and the use of a parabolic fractal plot developed by Laherrère of field size versus rank, and examining the

development of this plot over time. Estimates of oil yet-in-place were also evaluated in the light of geological knowledge, and converted into yet-to-find based on assumptions of recovery factor and drilling capacity.

- For each country addition of its cumulative production, P50 reserves and yet-to-find gave an estimate of that country's ultimately recoverable reserves ('ultimate').
- Modelling each country's future production. For a country already past peak, this simply declined future production at the existing depletion rate (fixed percentage of the remaining recoverable resource). If prior to peak, the forecast increased production at an assumed growth rate until cumulative production equalled half that country's ultimate, and thereafter decreased production at the depletion rate at peak. In the case of the Middle East swing producers, production was calculated under a number of assumptions subject to each country's recoverable resource limits.

These Petroconsultants studies led to the key publication in *Scientific American* in 1998, *The End of Cheap Oil*. See also details of Reading University's learning from Petroconsultants in Chap. 4 of Campbell (Ed.) (2011); and the comments on the derivation of the Petroconsultants data, and the company's subsequent buy-out by IHS Energy, in Sect. 9.4 of Hall and Ramírez-Pascualli (2013).

A5.6 Deffeyes

Deffeyes knew Hubbert, and authored two excellent books explaining the global oil peak, primarily using the Hubbert linearization technique applied to global data.

As mentioned earlier, such linearisation can be very powerful, but usually only works well once a region is close to or past peak (i.e., once the production curve has established a logistic shape; see the many examples in Campbell's *Atlas*, 2013). Indeed, Laherrère writes: "*The best way is to estimate the ultimate reserves (area below the annual production from start to end) from the 'creaming curve', by extrapolating the cumulative backdated 2P discoveries versus the cumulative number of fields (or of new field wildcats).*" Linearisation of production should only be used "*when* [data on] *backdated discoveries are not available*".

It is true that Deffeyes did not have access to an industry 2P database, and hence his calculations were necessarily approximate. Nevertheless, they provided clear insights to the forthcoming global oil peak.

[Incidentally, on Deffeyes' forecast date for global peak, Craig Bond Hatfield wrote (see Endnote 12) : "*I think that Deffeyes' prediction of Thanksgiving Day, 2005, as the time of maximum world oil production rate was a playful reflection of his sense of humor and was meant to draw attention to the fact that no one can know precisely when oil production rate will peak or precisely when it will begins its long-term decline. He probably thought that an obviously facetious forecast, far more precise than available data can permit, might at least draw some attention to the problem and its urgency.*"]

A5.7 Globalshift Ltd

Dr. Michael Smith of Globalshift has recently generated a new detailed 'bottom-up' forecast model of global oil and gas production, with results at: www.globalshift.co.uk.

For every oil-producing country there is a graph of past and forecast oil and gas production from 1950 to 2030; data on exploratory and development wells drilled from 1980 to 2010; and a short description of the country characteristics. Underlying the graphs are Excel spreadsheets including data back to when production and drilling began, and forecasts to 2100. The graphs are fascinating; see for example those for the US, Brazil, Russia, Nigeria, Iraq, Iran and Saudi Arabia.

Global production is also shown from 1950 to 2030. The model calculates that global production of fossil oil ('oil refined from on and offshore reservoirs, including tight sands and shales, and NGLs') will peak about 2025 at ~115 Mb/d; As the site explains, the actual date of peak will depend strongly on demand and other issues related to price, and on how the 'swing producers' behave.

Key of course are the data and methodology used to create these forecasts. Some information is on the website; see also the entry for the earlier 'Energyfiles' model in the UKERC study.

Dr. Smith states that: "*The model is not driven primarily by field reserves (which are subject to revision), but by reasonable expectations of field production based on past performance, plus allowance for new finds and technology gain. The model assumes that the oil price will, on average, respond to demand—which will be strong but volatile, so most fields, including most of those currently fallow and undiscovered, will eventually come on-stream.*" As Smith points out, the website is about providing the wider energy community with oil and gas forecast data based on a consistent and comprehensive technical analysis.

For a recent description of Globalshift Ltd.'s 'all-liquids' forecast model, see Smith (2015).

A5.8 Peak Oil Consulting

This consulting company is led by Chris Skrebowski who for many years was Editor of the Energy Institute's *Petroleum Review*, and ran a global 'mega-fields' model. The latter has subsequently been expanded to forecast all production. It includes mainstream oils and tar sands, but not CTL, GTL and biofuels. The model recently forecast that oil demand, if it were to grow by 1.6 % pa, would outstrip global production capacity around 2016. Skrebowski has also produced important graphs of the main oil exporting countries, showing declining scope for exports as home consumption grows (a point also made by Mitchell at Chatham House).

Peak Oil Consulting's data were used in the reports by the UK's Industry Taskforce on Peak Oil & Energy Security (ITPOES). This organisation consists of a number of major companies concerned about security of future hydrocarbon supply. (See www.peakoilconsulting.com; and also www.peakoiltaskforce.net)

A5.9 Other Mid-Point Peaking Models Based on an Assumed URR

A number of mid-point peaking oil forecast models have been produced based on assumed values for the global URR, including that from the German BGR, and the consultancy LBST.

A5.10 Uppsala University

One of the best sources for academic papers on topics related to all fossil fuel limits, including peak oil, is the Global Energy Systems Group in the Department of Earth Sciences (formerly in the Department of Physics) at Uppsala University, Sweden.

They have published many peer-reviewed papers. Topics cover oil supply (including an important explanation for the Hubbert curve based on economic considerations), gas supply (including forecast Norwegian & Russian production), coal, and fossil fuel CO_2 emissions compared to the standard IPCC SRES scenarios.

Professor Aleklett, now Emeritus, originally headed up this group, and wrote the excellent reference book, *Peeking at Peak Oil*, referred to earlier (Aleklett 2012), which will soon be re-issued in an update. Professor Mikael Höök is the Group's current Head.

A5.11 UKERC

As mentioned above, in 2009 the UK Energy Research Centre (UKERC) published a comprehensive study of analyses of future oil supply: S. Sorrell *et al.*, *Global Oil Depletion*, UKERC, 2009. (From: www.ukerc.ac.uk, via: Research/Technology & Policy Assessment.)

This concluded that "*On current evidence ... a peak of [global] conventional oil production before 2030 must be considered likely*"; and furthermore "*there is a significant risk of* [this peak] *before 2020.*"

The main report was backed up by seven technical reports:

1. Data sources and issues
2. Definition and interpretation of reserve estimates
3. Nature and importance of reserve growth
4. Decline rates and depletion rates
5. Methods of estimating ultimately recoverable resources
6. Methods of forecasting future oil supply
7. Comparison of global supply forecasts

As mentioned in Chap. 4, the Annex: *Models of Global Oil Supply for the Period 2008–2030* to Technical Report 7 examined forecasts from the IEA, OPEC, US EIA, BGR, Shell, Meling (a private forecast from a senior employee of StatoilHydro), Total, ExxonMobil, Energyfiles, LBST, Peak Oil Consulting, Colin Campbell, the University of Uppsala, and Richard Miller.

A5.12 The IEA

The story of the IEA's *World Energy Outlook* is on the whole disappointing. This is particularly so as many people in governments and other organisations have regarded the IEA *WEOs* as "the bible" on future oil supply.

The earliest *World Energy Outlook* followed a general econometric approach, and was apparently unaware of the 'technically-based' studies on future oil supply available at the time that predicted the world conventional oil peak as around the year 2000 (see Chap. 3). Then, as explained previously, when the oil price fell dramatically in the mid-1980s the popular conception of oil 'running out' came to be seen as erroneous, and oil analyst opinion swung almost completely to the model of 'reserves are just inventory', and the total quantity of oil available was "unknown and unknowable".

The IEA apparently became convinced of the validity of this approach, and for over a decade, from about the mid-1980s to mid-1990s, *WEO's* forecast *demand* in detail by sector, but were reassured by the *total quantity* of global recoverable oil expected to be available, and hence did not model supply in detail, asserting that there was 'more than enough oil available to meet demand' over the forecast period.[17] This ignoring of the 'mid-point peak' for conventional oil constituted a major error.

This view changed when J-M Bourdaire joined the IEA. He organised a key meeting on oil supply at the IEA in 1997; see recollections of this meeting by several of those present in Campbell (Ed.) (2011). As an outcome of this meeting Bourdaire oversaw the publication of the 1998 *WEO*. This was explicit about the expected global peak of *conventional* oil. It used as base case the then-USGS value for mean global ultimate for conventional oil, of 2300 Gb, and predicted that the global peak of all-oil (except for 'unidentified unconventional oil') would occur in 2014. Sources since have made it clear that choice of the phrase 'unidentified unconventional oil' was indeed intended to highlight the expectation of an all-oil peak.[18, 19]

Subsequently, Bourdaire left, and also at about this time the USGS produced its year-2000 assessment of global undiscovered oil, which—as noted above—now included allowance for 'reserves growth' in regions outside the US. Though this approach had been discussed since the mid-1990s, and been discarded by Bourdaire and others on methodological grounds, it was accepted by the IEA's new Director, Olivier Appert, a former IFP Deputy Director-General and a believer in the capacity of science and technology to grow reserves.

Progress in understanding the global peak was subsequently made, however, when Dr. Birol who by now headed up the *WEO* team at the IEA and who had wanted for some time to study field and regional decline in detail, was now authorised to do so. A good study ensured, and the prospect of a possible near-term global oil peak naturally became apparent with the production of the fields in decline needing to be offset by production from recent fields and the yet-to-find.

Despite this realisation, the 'economic' mindset persisted both within the IEA, and widely outside, with *WEO*s in this period stating that though this decline from existing fields was large, and needed to be replaced, global supply would be adequate provided sufficient investment came forward. As a result, the concept of peak oil again got dropped in these next few *WEO*s, which once again largely only forecast demand in detail, and asserted that sufficient oil was available to meet the forecast.[20]

More recently at the IEA the focus on the need for investment alone has diminished, and above-ground issues, such as Iraq's potential, have been highlighted. Certainly, the IEA's modelling has become more realistic over time; but it is a great pity that its flawed earlier studies were so widely treated as authoritative (and see Fig. 5.2).

We can also look specifically at the IEA (and also US EIA, see next section), forecasts for Saudi Arabian production, as reported in the UKERC 2009 *Global Oil Depletion* report; Fig. A5.1.

This figure is taken from *Technical Report 7* of the report, and shows the results from five forecasts made in 2008 for Saudi Arabian output to 2030. As can be seen, two of the forecasts (from the IEA and the US EIA) showed no peak in production before 2030, the other three indicated a peak. We can look at this more closely by examining the forecasts that indicate peak. The Campbell forecast was based on the assumption that Saudi Arabian reserves are significantly overstated. Miller's forecast was bottom-up by field, and includes his view of Saudi Arabian reserves, yet-to-find and likely reserves growth. It shows the 'Miller bump' in production that would occur if the country's fallow fields were brought on-stream almost immediately. Miller shows this case, but reports that he thinks it very unlikely. Energyfiles' forecast was also bottom-up by field, but seemed to take a more pessimistic view of the fallow fields production. On the basis, one can see that the IEA and US EIA forecasts at that date were not impossible given the oil present in the fallow fields, but both would have required—if the evaluations of Miller and

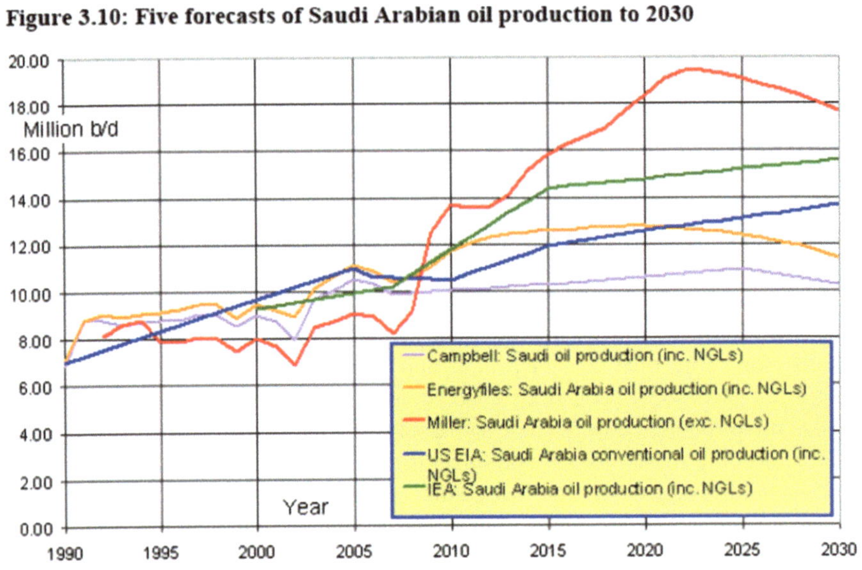

Figure A5.1 Five forecasts made in 2008 of Saudi Arabian production to 2030.
Source Technical Report 7 of UKERC *Global Oil Depletion* study, 2009

Energyfiles of available oil are roughly correct—that total Saudi Arabian production would peak soon after 2030.

For the IEA's fairly recent view of future global oil production, see Fig. 4.4. Additional references on IEA modelling include:

- On past IEA forecasts: Bentley and Boyle (2008).
- On the IEA's more recent modelling in *WEO*s, see:
  - K. Aleklett et al. *The Peak of the Oil Age—analyzing the world oil production Reference Scenario in World Energy Outlook 2008.* Energy Policy, vol. 38, 3, pp 1398–1414; 2010.
  - Miller, R. G., *Future oil supply:The changing stance of the International Energy Agency.* Energy Policy, vol. 39, pp 1569–1574; 2011.
  - Section 9.3 of Hall and Ramírez-Pascualli (2013).

A5.13 US EIA

The US' EIA have a long history of forecasting future oil supply, but are generally recognised as having made a pretty poor fist of it. This is partly understandable. The US' over a century of poor proved reserves data led to both a fixation on an R/P ratio of 10, and a misleading view of the importance of reserves growth. And, as noted above, later probably too much reliance was placed on USGS 'assessed' estimates of undiscovered oil, rather than on 2P discovery trends. As a result, like the IEA, the EIA for a long time simply modelled demand and assumed that supply would be adequate, ignoring the mechanism that drives peak.

I am not conversant with the current EIA model, but the EIA's formal reply a little while back to a sensible letter from ASPO-USA did not give confidence.

(Incidentally, also within the US there had been a concerted effort to look at future oil and gas supplies under the auspices of the National Petroleum Council. The resulting report (NPC 2007): *Facing the Hard Truths about Energy* raised concerns about future supply, but did not seem to properly understand peak.)

A5.14 Oil Company Views

This section covers the changing views from oil companies. Because of the influence of BP, at least in the UK, their views are covered at greater length.

The general picture of the oil companies' views is as follows:

- As reported in Chap. 3, in the early-1970s to mid-1980s, from before and following the oil shocks of the 1970s, a number of the oil majors carried out sensible 'technically-based' modelling, and warned that the global peak or plateau of conventional oil production (ex-NGLs) would occur around the year 2000.
- With the fall of the oil price in the mid-1980s, interest in such modelling largely fell away.
- Subsequently, in the absence of this modelling, and with perhaps the rise of economists as company spokesmen, the view turned to one of cornucopia.
- Today this 'cornucopia' view can still be heard (see, e.g., BP and Chevron, below), but recently more attention is being paid to oil forecast models which are now quite good in some of the majors; and also now acceptance is sometimes publically expressed that that the world has reached 'the end of its easy oil'.

The above 'trajectory of view' therefore has some parallels to that at both the IEA and US EIA, as described above. Moreover, the public are often very surprised to learn that for a long while only relatively little modelling of global supply (or in some cases, none) was being done by the majors.[21] But it is also important to point out that—unlike the IEA, or the EIA—it was *not the oil majors' job* to analyse global supply and warn the world of future oil shocks; this was mentioned explicitly on a number of occasions to me by oil company personnel. Moreover, for many years the majors' main interest in the future oil price lay in risk that downward trends in price might affect the business; possible upside price trends were seen simply as a boon.

Now we turn to forecasts by specific oil companies:

A5.14.1 Shell Oil Company

As mentioned in Chap. 3, in 1979 Shell forecast global oil production to "plateau within the next 25 years." At that time, Shell had already become well known for their scenario modelling, which had indeed earlier warned the company—much against the general expectation of the time—of the possible oil price shock in prospect. (See Jefferson 2012, for a good discussion of Shell's scenario building over a number of years.) This view of approaching global peak persisted, possibly as late as mid-1990s. Certainly, still at this date Hans DuMoulin, who had been at Shell, explained to an ETE 21 meeting in London that we attended about the global oil production 'mountain', with its peak and inexorable descent on the other side.

But when, a little later, two representatives of Reading University 'Oil group' went to Shell's London HQ in the late-1990s to discuss the possibility of a global oil peak, we were told this was just a phantom: Shell's detailed global model showed that more than enough oil would be available to meet global demand for the foreseeable future.

However, today we know from fairly recent 'Oil Forecasting' workshops held by the UK's Energy Institute in London, that Shell has now a detailed and sensible global oil supply model.

A5.14.2 ExxonMobil

As also mentioned in Chap. 3, in 1972 ESSO quoted a global conventional oil ultimate (almost certainly excluding NGLs) of 2100 Gb, and hence said they expected oil to become "*increasingly scarce from about the year 2000*".

Exxon later moved away from this position, and by 2001 told the UK House of Lords Select Committee (2002) looking at energy security: "*Resource availability is simply not the issue.*" [22]

Exxon's most visible retraction of their 1972 position came in their widespread advertising campaign in 2006 (when the oil price was already on the rise, and Campbell's 'Regular conventional' oil already in decline) that was headlined: "*No peak in sight*", and which said that such an oil peak: " *... will not occur this year, next year or for decades to come*".

Exxon's subsequent position in 2008, within the UKERC *Global Oil Depletion* study, has been reported above in Fig. 3.1 in Chap. 3, and their current view in Fig. 4.6.

A5.14.3 Chevron

By contrast to ExxonMobil, at about the same date as Exxon's advertisement Chevron were running a series of advertisements warning—if slightly elliptically—of global oil supply difficulties ahead. It may well be that the company had come to understand this issue through an excellent study carried out by PFC Energy, who by then (though not before!) were incorporating field decline into their oil forecast models.

Subsequently, however, Chevron's later 'Will you join us' advertisements (incidentally, themselves a model of public involvement) no longer even hinted at oil supply difficulties, but instead referred to other important energy and environment issues. This change of view was reflected in a speech made by the CEO of Chevron, perhaps three years' back at a UK Energy Institute's 'IP Week', which dismissed any fear of global oil supply difficulties on the basis that "*world proved reserves continue to increase*."

A5.14.4 ENI

A few years back ENI's Chief Economist expressed the 'oil cornucopia' view in a 'Comment' section in *Science* (Maugeri 2004). This was rebutted in Meng and Bentley (2008).

A5.14.5 Total

Among the oil majors, for a while it seemed as if only the public statements by Total's Chief Executive, Christophe de Margerie, clearly recognised the likelihood of relatively near-term of oil peaking.

But now there seems to have been a change of view within the company, with de Margerie saying more recently that 'new sources of petroleum, such as tight gas and shale oil, mean that the world had ample supplies of petroleum', and that 'while there were economic and environmental issues which would affect how quickly resources were exploited, there was "*definitely not a concern about reserves*"'.

A5.14.6 BP

Like the other oil companies discussed above, BP's publically expressed view on the likely timing of a world oil production peak has changed significantly over time, as follows:

(a) 1979 report by H.R. Warman

In 1979 Harry Warman of BP produced a report titled *Oil Crisis ... again?* which forecast future oil production, almost certainly based on a 'mid-point' approach (Warman 1979). The report predicted that world oil production, outside communist areas and excluding NGLs, would peak about 1985 This report was widely cited later by critics such as Peter Odell, John Mitchell, and Vaclav Smil as proof of the inability of 'fixed resource' models to forecast oil production.

Examination of Warman's data, however, reveals that his estimate for the size of the global non-communist ex-NGL recoverable oil resource was reasonable. The problem, just as with Hubbert's prediction for the global peak in 1996, was that Warman's 1979 report also assumed unfettered global production growth toward peak, and did not reflect the price-driven demand reductions of the 1980s that subsequently occurred.

(b) Calculations by Harper and Miller

Within BP, as with nearly all other modellers, with the fall in oil price in the mid-80s the focus on potential global oil supply limits largely evaporated. For a while it may be that the only quantitative global resource analysis within the company was that by Francis Harper; and the only detailed forecasting that by Dr. Richard Miller. The latter had generated global a bottom-up, by-field forecasting model, the results of which were circulated to some of the senior management.

Because the views of Harper and Miller conflicted starkly with the 'no peak' view of the company's Chief Economist, Peter Davies, a high-level company meeting was called, which decided that control of what was to be said externally on oil supply should be assigned to Davies.

(c) View of Peter Davies

This brings us to the widely influential views of Davies. As mentioned in Chap. 5, Davies' position on future oil supply reflected the prevailing economic view that 'reserves were just inventory'; that experience (based on proved reserves data) had shown that these reserves were continually being replaced over time; and that—in any event—there could be no risk of any *near-term* oil peak because of the large size of global proved reserves, standing at over forty years' of secure supply.

When giving oral evidence to the House of Lords Select Committee mentioned above, Davies said: "*As of today there are very adequate supplies of oil and gas. With the proven reserves and today's technology there is 40 years of oil and 60 years of gas plus*" (UK House of Lords 2002, p79). And when questioned by the committee on ODAC's warning of a proximate peak in global production, his reply was dismissive and believed.[23]

Subsequently, at a meeting of the UK's All-Party Parliamentary Group on Peak Oil in the House of Commons Davies reasserted that any global peak had to be far off because: "the world has 40 years' of proven reserves."

(Note that members of the 'Oil group' at Reading University tried on a number of occasions to discuss this view with Davies, but to no effect.)

(d) Subsequent views from BP

Unfortunately, within BP this error about the security of global oil supply, based on simply quoting the current size of the proved reserves, has been persistent. In June 2008, for example, Tony Hayward, then BP's CEO, said: "Myth number two is that the world is running out of hydrocarbons. Not so. The world has ample resources, with more than 40 years of proven oil reserves …".

And as mentioned earlier, BP's recent Chief Economist, Christof Rühl, blogged in 2014: " ... *all the global "end of oil" predictions so far have been wrong. Proved reserves, for example, more than doubled since 1980—despite the world consuming more oil over the last 34 years than it had in proved reserves in 1980.*" (Rühl 2014).

The published views of the company's current Chief Economist remain very similar.

(e) BP's current modelling

Despite this, however, things have been changing within the company. In recent years, the company has again started to officially run its own internal detailed oil forecasting model, and the company has released results from this for the last few years, see www.bp.com. In the *BP Energy Outlook 2030* issued in January 2012, the company forecast global all-liquids supply to grow modestly from 87 Mb/d in 2010 to 103 Mb/d by 2030, which—for all-liquids—is certainly plausible. The company's latest global forecast is given in Fig. 4.5.

(f) Subsequent modelling by Miller

After retiring from BP, Miller has continued to run his model with new data, and its details and results as of 2008 were covered in the UKERC (2009) report.

Subsequently, as mentioned earlier, Miller has expressed scepticism of the rate that fallow fields were brought on-stream in that version of the model, and today suggests that a combination of below-ground and above-ground factors limit global production to a 'bumpy plateau' at around current production levels. He writes: "*My current sense is that we have reached an effective price-constrained plateau. This price may rise gently with time as people find ways to divert money to buying energy, and as money itself inflates. I do not think that there is enough quickly-developable, cheap, non-conventional oil to make a difference*". For a recent detailed description of his 'all-oil' forecast model, see Miller (2015).

In addition, useful work has been undertaken at by Christophe McGlade at UCL to expand Miller's 'unconstrained' model into one that incorporates scenarios of demand, takes account of production cost (so cheaper fields are developed before more expensive ones), and allows for substitutes from other sources. For a description of this see McGlade (2015).

A5.15 Banks and Investment Houses

Finally, in this Annex on oil forecasts, we note that forecasts from organisations such as banks and investment houses, though widely quoted, have usually been poor or very poor (with some rare exceptions, such as that from the CIBC). In part, this has been because the models created by such organisations usually have little or no input from either petroleum geologists or reservoir engineers, the key disciplines required to assess the world's quantity of accessible oil as a function of price.

Annex 6
Peak Oil and Climate Change

In this Annex, we look at the important connection between peak oil and climate change. The links between these two topics are complex, and would need much discussion to cover properly. Here we only point out briefly some of the issues, and give some references.

(a) Issues

The two topics have similarities. For example, some of the responses required—such as the need to reduce use of oil and increase efficiency of use—are the same; and also both topics require better understanding by the public, and by some in government.

But there are key differences: peak oil has already affected society and will do so increasingly, whereas the almost-certain very significant impacts from climate change are mostly further in the future; also peak oil will cause social disruption (and almost certainly conflict if not handled well), whereas climate change is the more serious because if not handled well it threatens—potentially—the existence of mankind.

So far the two 'camps' (the oil peakists and the climate change scientists) have had relatively little interaction. Some oil peakists, being geologists, see long-term climate change as having occurred often in the past and as inevitable, so are sceptical over the significance of short-term changes; while the climate scientists say—correctly—that there is too much oil, and much must be left in the ground, so why be concerned with peak oil?

The answer is simple: it is the world's energy use that drives anthropogenic climate change, and both topics must be properly understood to have a correct view of mankind's future.

(b) Sources and References:

1. The first organisation known to me to quantitatively link peak oil to climate change was Global Commons, in a presentation on their widely accepted 'contract & converge' principle. In this presentation they showed, on the data then available, that peak [conventional] oil and peak [conventional] gas would contribute adequately to meeting the desired global carbon

R.W. Bentley, *Introduction to Peak Oil*, Lecture Notes in Energy 34,
DOI 10.1007/978-3-319-26372-4

pathway over time, but that significant action would be needed to limit coal emissions if this pathway was to be met.

2. Reading University's 'Oil group' made a presentation to the UK's Royal Commission on Environmental Pollution on peak oil when the latter was taking evidence on climate change. Reading presented data supporting a near-term oil peak, and Peter Odell a very much later one, around 2060. The Commission's view was that whether the peak was early or late was of no consequence, as oil use had to be curtailed in any event (specifically: "*keep the stuff in the ground*").
   This view was correct, but took no account of the impact of peak oil or gas on likely future emissions, nor of the probable difficulty of implementing climate change mitigation and adaptation measures in an economically constrained post-oil-peak world. [I think this input by Reading was to the RCEP's 22nd report (2000), *Energy—the Changing Climate*, which—again I think—was the first document to recommend a 20 % reduction in UK CO_2 emissions by 2010 and a 60 % reduction by 2050, based on the evidence then available that atmospheric CO_2 concentrations needed to be maintained below 550 ppmv to avoid "dangerous and destructive climate change".]
3. Perhaps the earliest journal publication linking the two topics was that by Michael Grubb (2001). This article, '*Who's afraid of atmospheric stabilisation?*', analysed the link between energy resources and climate change, and showed (if memory serves), rather like the Global Commons article, that emissions from *conventional* oil plus *conventional* gas would not breach the then-suggested global carbon budget, but that significant production of non-conventional sources of these fuels would; and hence that this provided a possible, politically-acceptable, avenue for focussed CO_2 permits.
4. P.A. Kharecha and J. E. Hansen (2008) also looked at the linkage in '*Implications of "peak oil" for atmospheric CO_2 and climate.*' This concluded that:

   > *If conventional oil production peaks within the next few decades, it may have a large effect on future atmospheric CO_2 and climate change, depending upon subsequent energy choices. Assuming that proven oil and gas reserves do not greatly exceed estimates of the Energy Information Administration ... we show that it is feasible to keep atmospheric CO_2 from exceeding about 450 ppm by 2100, provided that emissions from coal, unconventional fossil fuels, and land use are constrained.*

5. Jeremy Leggett covered the confluence of the two topics in *Half Gone* (Portobello Books, London 2005); and also in a later book (*The Energy of Nations—Risk Blindness and the Road to Renaissance*, Routledge 2014) which included the stresses the two topics would place on the global economic system.

6. Jean Laherrère has for a long time published graphs showing how the IIASA high-end SRES emissions scenarios originally used by the IPCC were not compatible with either peak oil or peak gas; and Kjell Aleklett in his book, *Peeking at Peak Oil* (Springer 2012) has a chapter looking at this topic, including his discussion directly with IIASA on this. SRES scenarios have been superseded by Representative Concentration Pathways, but these owe much to the original SRES thinking; and again the high-end values look unrealistic under likely peak oil or peak gas scenarios.
7. Similar views are given in Ward et al. (2012), and in Höök and Tang (2013); while a detailed view of the match between climate change emissions assumptions and a range of published supply-side analyses of the depletion of all three of the main fossil fuels is given in Wang et al. (submitted for publication).
8. Finally, Christophe McGlade and Paul Ekins, of University College's Institute for Sustainable Resources, have looked at the topic in: *Un-burnable oil: An examination of oil resource utilisation in a decarbonised energy system*, by published in an *Energy Policy* Special issue: *Oil and gas perspectives in the 21st century,* Elsevier, 2014. (Note that this paper was written prior to the latest IPCC-5 reports, where the latter yet-again strengthened the case for the anthropogenic origin of current climate change.) They revisited the topic in a widely discussed letter to *Nature* titled: *The geographical distribution of fossil fuels unused when limiting global warming to 2 °C.*

(c) Summary

There is no doubt that the confluence of the two topics: peak conventional oil production (and, as well, peak conventional gas, and possibly also peak cheap hard coal) and climate change needs to be properly understood; and that overall global use of oil and gas, probably of both conventional plus the non-conventionals, will need to be restrained if the 2 °C limit is to be avoided. It is for this reason that the start of Chap. 1 said: "This transition [caused by resource limits to conventional oil supply] … will probably … lead to an overall decline in global oil supply". As with 'peak demand' below, measures to avoid serious climate change may turn out to be a stronger reason for global oil supply to fall.

Annex 7
Peak Demand

This Annex discusses 'peak demand'. Note that this term refers to two very different phenomena, for which here we coin the terms 'supply-driven' and 'exogenous' peak demand:

- 'Supply-driven' peak demand: Since oil supply and demand must match, if there are limits to supply then demand must also be limited (often by a rise in oil price, though sometimes through rationing or similar). This limiting of demand by price is almost certainly happening currently in many OECD countries, and would likely happen in a wider range of non-OECD counties if the oil price were somewhat higher (or equivalently, in some countries, if the subsidies were lower).
- 'Exogenous' peak demand: By contrast, a second type of oil 'peak demand' is where demand for oil declines for reasons *not directly related to* the current availability of supply. Such a peak in demand would result, for example, from increases in vehicle efficiency or greater use of natural gas and electric vehicles, where these are driven by climate change or future energy security concerns, or by other societal reasons *not directly linked* to the *current* availability of oil.

Of course both drivers (a supply limit / high price, and demand changes due other factors) may well act in combination.

Several groups currently make predictions of fairly near-term global 'peak demand', for example Shell, Deutsche Bank, Citigroup and Ricardo, and see this peak as resulting from different combinations of the above factors. But it may be that none of these groups fully understands the *resource-limited* driver for oil's current high price as set out in this book, nor the coming peak in the production of conventional gas, so their 'peak-demand' predictions miss these crucial 'supply-side' aspects of the problem.

This is borne out, for example, by a Ricardo Strategic Consulting press release of 7th November 2011, which said:

R.W. Bentley, *Introduction to Peak Oil*, Lecture Notes in Energy 34,
DOI 10.1007/978-3-319-26372-4

Ricardo study suggests global oil demand may peak before 2020.

- *Research challenges the concept that 'Peak Oil' will be a supply-side phenomenon.*
- *Instead, demand for oil may well peak before 2020, falling back to levels significantly below 2010 demand by 2035. Significant changes in future demand patterns [will be] strongly influenced by global energy security policies, the technology change that they promote, and demographics.*
- *Evolutionary change in automotive technology is predicted to bring revolutionary change in fuel demand.*

"The world is nearing a paradigm shift in oil demand," explains Peter Hughes …

- *As a result, the drivers working against oil demand growth are increasing in number and intensity, with the world's consuming nations increasingly focussed on their need to reduce their dependency on oil, supported by an ever stronger legislative framework.*
- *… suggest that efficiency improvements in the internal combustion engine will more than offset the rise in fuel demand deriving from the increase in the number of vehicles.*
- *… If crop yields increase at historic rates, there will be enough surplus conventional fuel crops to displace a significant amount of fossil fuels.*
- *… the improving supply outlook for natural gas, with the potential for the surge in shale gas production in North America to be replicated elsewhere over time … [is likely to encourage] substitution of oil in both stationary and on-road transportation."*

As mentioned above, this study—while very worthwhile in itself—almost certainly did not properly factor-in the data on peak conventional oil (or peak conventional gas) as set out elsewhere in this book.

Annex 8
Peak Gas

As Fig. 2.22 shows, on a simple 'mid-point' view, the peak of global *conventional* gas production is expected about 2025 unless there are very major new conventional gas finds. (Also note that the reserves of conventional gas are even more geographically restricted than with oil, being mainly in Russia, Iran and Qatar.)

Like oil, there are large quantities of non-conventional gas in the world, including shale gas (see Table 3.1), other tight gas, gas in deep brines, gas from coal bed gasification, and—possibly—from methane hydrates.[24] But as with oil, it is almost certain that the peak of conventional gas will have a significant effect on the gas price, and possibly on overall gas supply also. For 'all-gas', Campbell, for example, forecasts the global peak at around 2020 (where this includes infrastructure limits); while Smith of Globalshift Ltd. forecasts the all-gas peak, excluding gas from underground coal combustion (UCG) and the methane hydrates, by 2030.

R.W. Bentley, *Introduction to Peak Oil*, Lecture Notes in Energy 34,
DOI 10.1007/978-3-319-26372-4

Annex 9
Reading University 'Oil group'

Past and present membership of the UK's University of Reading very *ad hoc* 'Oil Group' has been:

Postgraduate Research Institute for Sedimentology:
Prof. M.L. Coleman (ex-BP), Prof. B.W. Sellwood.
Department of Engineering:
Dr. J.D. Burton, Mr. R.H. Booth (ex-Shell),
Dr. R.M. Mayer (ex-BP), Prof. P.D. Dunn;
MSc. Students: L. Kurdgelashvili, S. Mannan, S. Wheeler.
Department of Cybernetics:
Dr. G.R. Whitfield, Dr. R.W. Bentley (ex-Imperial Oil & Esso Chemicals).
Affiliated: Dr. D. Fleming, independent economist.
Also Students: S. Olayiwola, City University; Q. Meng, CUP, Beijing, China.

For many years, as far as we know, this was the only academic group in the UK carrying out quantitative study of future global hydrocarbon supply. Publications by members of the group on global oil supply, in addition to those listed elsewhere in this paper (see References), are:

2012: S. Sorrell, J. Speirs, R. Bentley, R. Miller, E. Thompson. *Shaping the Global Oil Peak: A review of the evidence on field sizes, reserve growth, decline rates and depletion rates*. Energy. Elsevier, vol. 37, Issue 1, January 2012, pp 709–724.

2010: S. Sorrell, R. Miller, R. Bentley, J. Speirs. *Oil futures: A comparison of global supply forecasts*. Energy Policy, vol. 38, pp 4990–5003, 2010.
S. Sorrell, J. Speirs, R. Bentley, A. Brandt, R. Miller. *Global oil depletion: A review of the evidence*. Energy Policy, vol. 38, pp 5290–5295, 2010.
R.W. Bentley. *The Expected Dates of Resource-limited Maxima in the Global Production of Oil and Gas*. Energy Efficiency, vol. 3, pp 115–122, 2010.

2009: R.W. Bentley. *An explanation of oil peaking*.—Manuscript, 2009.

2003: R.W. Bentley and M.R. Smith. *World Oil Production Peak—A Supply-Side Perspective*. Presented at Intl. Assocn. for Energy Economics Conference, Prague, June 2003.

R.W. Bentley, *Introduction to Peak Oil*, Lecture Notes in Energy 34,
DOI 10.1007/978-3-319-26372-4

2002: R.W. Bentley. *Global oil and gas depletion: an overview.* Energy Policy, vol. 30, no. 3, February 2002, pp 189–205. Elsevier, 2002.
R.W. Bentley. *Oil Forecasts, Past and Present.* Energy Exploration and Exploitation, vol. 20, no. 6, pp 481–492, Multi-Science Publishing Co. Ltd., 2002.
(Paper originally given at the First International Workshop on Oil Depletion, IWOOD2002, Uppsala University, Sweden, May 23–24, 2002; see website: www.peakoil.net; proceedings of IWOOD2002).
2000: R.W. Bentley, R.H. Booth, J.D. Burton, M.L. Coleman, B.W. Sellwood, G. R. Whitfield. *Perspectives on the Future of Oil.* Energy Exploration and Exploitation, vol. 18, nos. 2 & 3, pp 147–206, Multi-Science Publishing Co. Ltd., 2000.
R.W. Bentley. *Were those past oil forecasts really so wrong?* Published by OneOffshore Ltd., part of the PetroData Group, in the industry newsletter: *Tomorrow's Oil*, November, 2000.
R.W. Bentley, R.H. Booth, J.D. Burton, M.L. Coleman, B.W. Sellwood, and G.R. Whitfield. *World Oil Supply: Near & Medium-Term.* Report submitted to the DTI's Energy Advisory Panel, Sept 5th, 2000.
1999: R.W. Bentley, R.H. Booth, J.D. Burton, B.W. Sellwood, G.R. Whitfield. *The Oil Future—A Very Different View.* Newsletter, Int'l. Assocn. for Energy Economics, 4th quarter, 1999.
1998: R.W. Bentley. *UK Energy—The Next 5 to 10 Years.* Report submitted to the UK Dept. of Trade & Industry, and circulated among UK and overseas energy analysts. Jan. 1998. Updated in 1999 as: *Oil—The Next 5 to 10 Years*, and subsequently expanded as: *Perspectives on the Future of Oil.*
1997: R.W. Bentley. *Oil Shock 'Imminent' if Heavies are Slow or Expensive to Produce.* Energy World, June 1997, pp 20–22. The Institute of Energy, London.
1996: R.W. Bentley. *The Conventional Oil Production Peak—5 or 20 Years Away?* Energy World, Dec. 1996, pp 11–12. The Institute of Energy, London.

End Notes

1. *Cost of Shale Oil in the US*

While the cost of producing 'light-tight' oil in the US is usually reckoned to be of the order of $60/bbl; the long period of low oil price in 2015 seems to have encouraged some suppliers (especially those at the more favourable sites) to be able to lower this cost. A Bloomberg news article in August 2015 said:

Bloomberg 12/8/15: '$30 oil no problem for some Bakken drillers':

The breakeven price, based on production rates and drilling, completion and other costs, can vary widely within a play based on how prolific the geology is and the efficiency of the drilling company, according to Bloomberg Intelligence energy analysts William Foiles and Andrew Cosgrove.

In McKenzie County, North Dakota, one of the core areas of the Bakken, the median breakeven price is a little more than $29 a barrel, Foiles said. That's about a third less than in nearby Williams County, and it's less than half the average breakeven price for the Bakken that banks and research firms estimated last fall. McKenzie County wells have shown the best returns amid the price drop. Drillers had 26 horizontal wells seeking oil in that county last week, the most in the state, according to Baker Hughes Inc.

Bakken oil production in North Dakota has fallen less than 2 % from its peak in December, while the number of oil rigs in the state has fallen by 60 %. EOG Resources Inc., the largest shale driller, says it can make a 30 % after-tax return on $50 oil in its best plays. Whiting Petroleum Corp., the largest Bakken producer, said it's preparing to be able to grow production at $40 to $50 [/bbl] prices.

"A single break-even price doesn't actually exist," Foiles said in a presentation. "Rather, what the model indicates is that at a realized oil price of $29.42, half of wells will generate returns exceeding 10 %. This price is considerably lower than the $70 breakeven estimated by industry watchers at the start of the oil price slump."

Source: http://www.bloomberg.com/news/articles/2015-08-12/oil-at-30-is-no-problem-for-some-cost-cutting-bakken-drillers.

2. ***Economic Case for Synfuel***

I have not looked at the economics of synfuel in any detail. But one place where their economics may well make sense is for storage of surplus energy from an intermittent renewable energy source; such as from a wind array when more power is being produced than the grid requires.

3. ***Hubbert on Film***

M.K. Hubbert is on film explaining various aspects of peak oil in several informative clips on YouTube, and also in an excellent lecture made for the UK's Open University, introduced by L.F. Ivanhoe; see comment in Annex 5.

4. ***'Fiscal' Price needed for Oil by certain Exporters***

The 'fiscal price' of oil is the price that exporters need in order to make their government's current account balance meet zero. A government is not necessarily in serious trouble if the price of oil is below this; for example in 2015 to meet its deficit Saudi Arabia has drawn down its cash reserves, borrowed on its internal market, and planned reductions in government expenditure. But the 'fiscal' price gives an indication of where exporter governments would like to see the oil price if they are not to carry out such measures.

Estimated 'fiscal' oil prices for 2015 (forecast over the full year) are:

– Algeria	95 ($/bbl)
– Oman	85
– Iraq	69
– Saudi Arabia	68
– Bahrain	67
– Qatar	48
– UAE	44
– Iran	43
– Kuwait	39

- Forecast all-2015 average oil price: $56/bbl.
- By contrast, in 2012, though each countries' 'fiscal' oil prices were somewhat different from the 2015 values shown here, for all the above countries their 'fiscal' prices were well below the 2012 average oil price of ~ 110/bbl; i.e. these government budgets were all affordable in 2012.

Source: *The Economist*, Sept. 5th–11th, 2015, p 50; original sources given as: IMF, BP *Stats. Review*, Thomson Reuters

5. ***Useful 'Infographic' to Tie Oil Discovery to Production***

The concept of Fig. 2.4 could be used to enhance composite by-field graphs, such as Fig. 2.7, to visually tie discovery to production, and hence show that a specific region's production peak is indeed *resource-limited*, at least on the discovery data to-date. In such a graph for a region, each field's data would be indicated separately by colour or grey-scale, as in Fig. 2.7. Such a plot would then indicate each field's 'proved-plus-probable' discovery volume by a vertical bar at its discovery date, and also its production over time, stacked with other fields as is the case of Figs. 2.4 and 2.7. In such a plot the critical connection between the discovery of each field and its subsequent production would be clear, and hence also the production limit set by the discovered volume to-date. Plots of this type may exist, but I have not seen one, that I can recall.

6. ***Better Modelling of the Impact of Oil Price on Economic Activity***

It has always been disappointing that it would seem (at least to my knowledge) that nearly all studies examining the link between a high oil price and reduced economic activity have been based largely on broad correlation (e.g., oil price vs. change in GDP). There must be more than adequate data out there (for example, within company monthly internal business reports, government department budget documentation, and studies of individual spending) to be able to tie changes in oil price to specific decisions on economic activity made within these sectors. Such case-by-case analysis would help to explain—and hence significantly bolster—the validity of the data from broad correlations.

For example, in terms of the question of to what extent the high oil price was one of the triggers for the 2008 global recession, it has been reported that defaults on sub-prime mortgages in the US increased with distance of the mortgage payer from their place of work, and hence on their assumed cost burden for commuting. If it has not been done, it would be useful to back up such a finding with a survey of these mortgage payers to see to what extent the price of petrol did indeed affect their decision on whether or not to default.

7. ***The Significance of Approaching the Peak***

While Campbell, for example, is right to emphasise the significance of the decline of oil production post-peak, it was from a talk by Ron Oxburgh at a conference in Cambridge that I first heard the notion that it was not the peak of oil itself that was a key date, but the point where such an *approaching* peak forced oil supply to fall below the rate of increase in oil supply that a burgeoning world had come to see as normal ('business-as-usual'), and on which the world's expectations of economic growth are in part based.

8. ***The Zeitgeist of the late 1970s–early 1980s: 'Oil soon to run out'***

There are numerous examples in the 1970s of those who thought global oil would soon become in ever shorter supply. Here we give two influential examples, and also possible explanations underpinning these.

(a) *President Carter's 'Moral Equivalent of War' speech*

Perhaps the most notable example of this view was the 'Moral Equivalent of War' speech given by President Carter: *Address to the Nation on Proposed National Energy Policy*, on 18th April 1977. (Source: Wikisource, accessed 17th Sept 2014.)

> *Good evening.*
>
> *Tonight I want to have an unpleasant talk with you about a problem unprecedented in our history. With the exception of preventing war, this is the greatest challenge our country will face during our lifetimes. The energy crisis has not yet overwhelmed us, but it will if we do not act quickly.*
>
> *It is a problem we will not solve in the next few years, and it is likely to get progressively worse through the rest of this century. ...*
>
> *We simply must balance our demand for energy with our rapidly shrinking resources. By acting now, we can control our future instead of letting the future control us. ...*
>
> *The most important thing about these proposals is that the alternative may be a national catastrophe. Further delay can affect our strength and our power as a nation.*
>
> *Our decision about energy will test the character of the American people and the ability of the President and the Congress to govern. This difficult effort will be the "moral equivalent of war"—except that we will be uniting our efforts to build and not destroy.*
>
> *I know that some of you may doubt that we face real energy shortages. The 1973 gasoline lines are gone, and our homes are warm again. But our energy problem is worse tonight than it was in 1973. ... The oil and natural gas we rely on for 75 percent of our energy are*

running out. In spite of increased effort, domestic production has been dropping steadily at about six percent a year. Imports have doubled in the last five years. ..... Unless profound changes are made to lower oil consumption, we now believe that early in the 1980s the world will be demanding more oil that it can produce.

The world now uses about 60 million barrels of oil a day and demand increases each year about 5 percent. This means that just to stay even we need the production of a new Texas every year, an Alaskan North Slope every nine months, or a new Saudi Arabia every three years. Obviously, this cannot continue. ...

Because we are now running out of gas and oil, we must prepare quickly for a third change, to strict conservation and to the use of coal and permanent renewable energy sources, like solar power.

The world has not prepared for the future. During the 1950s, people used twice as much oil as during the 1940s. During the 1960s, we used twice as much as during the 1950s. And in each of those decades, more oil was consumed than in all of mankind's previous history.

World consumption of oil is still going up. If it were possible to keep it rising during the 1970s and 1980s by 5 percent a year as it has in the past, we could use up all the proven reserves of oil in the entire world by the end of the next decade. ...

I know that many of you have suspected that some supplies of oil and gas are being withheld. You may be right, but suspicions about oil companies cannot change the fact that we are running out of petroleum.

All of us have heard about the large oil fields on Alaska's North Slope. In a few years when the North Slope is producing fully, its total output will be just about equal to two years' increase in our nation's energy demand.

Each new inventory of world oil reserves has been more disturbing than the last. World oil production can probably keep going up for another six or eight years. But some time in the 1980s it can't go up much more. Demand will overtake production. We have no choice about that. ...

If we do not act, then by 1985 we will be using 33 percent more energy than we do today.

We can't substantially increase our domestic production, so we would need to import twice as much oil as we do now. Supplies will be uncertain. The cost will keep going up. Six years ago, we paid \$3.7 billion for imported oil. Last year we spent \$37 billion—nearly ten times as much—and this year we may spend over \$45 billion.

Unless we act, we will spend more than \$550 billion for imported oil by 1985—more than \$2500 a year for every man, woman, and child in America. ...

Now we have a choice. But if we wait, we will live in fear of embargoes. We could endanger our freedom as a sovereign nation to act in foreign affairs. Within ten years we would not be able to import enough oil—from any country, at any acceptable price. ...

These are the goals we set for 1985:

– Reduce the annual growth rate in our energy demand to less than two percent.

– Reduce gasoline consumption by ten percent below its current level.

– Cut in half the portion of United States oil which is imported, from a potential level of sixteen million barrels to six million barrels a day.

– Establish a strategic petroleum reserve of one billion barrels, more than six months' supply.

– Increase our coal production by about two thirds to more than one billion tons a year.

– Insulate 90 percent of American homes and all new buildings.

– Use solar energy in more than two and one-half million houses.

We will monitor our progress toward these goals year by year. Our plan will call for stricter conservation measures if we fall behind.

All-in-all, this was a sensible speech, and a sensible plan, given the zeitgeist that coupled the ~30 years' of global proved oil reserves with the concept that this was largely all the oil likely to be available. Crucially, the speech's underlying premise ("*... early in the 1980s the world will be demanding more oil that it can produce*") was apparently not informed by the forecasting work reported in Chap. 3. Nevertheless the plan led to many sensible steps, including research into renewables, and support for the CAFE automobile efficiency standards (first introduced in 1975) that—in ever-tightened form over the years—have stood the US in good stead.

In President Carter's case, there may have been an additional factor for this speech, in addition to the zeitgeist of the time. When at the ASPO-6 conference in Ireland I asked James Schlesinger, America's first Secretary of Energy and appointed by Carter, about the latter's views on oil, Schlesinger replied in effect: '*Always look to see who was a fellow's boss.*' Asked for clarification, Schlesinger pointed out that Carter had worked under Admiral Rickover. Since the latter had been Director of Naval Reactors in the US Navy, it seems likely that Rickover had impressed on Carter the Navy's vulnerability to interruptions in Middle East supply (and possibly also of the wider issues of future oil supply), and hence the need for the US to have a nuclear navy. This view of Carter's thinking is lent weight by Wikipedia reporting that: "*Carter later said that, next to his parents, Rickover had the greatest influence on him.*")

The situation envisaged by Carter—a near-term risk of declining global oil production—now looks likely today, despite significant production increases from the large resources of non-conventional oil. Hence an updated version of Carter's plan, and adapted for CO_2 concerns also, would seem warranted. In essence, Carter was not wrong on the plan, but wrong on its timing. (Incidentally on CO_2, it was under Schlesinger that the US DoE launched its innovative 'Carbon Dioxide Effects and Assessment' Program.)

(b) *The Energy Project at the Harvard Business School*

A second notable example of the 1970s zeitgeist on the likely future of oil is given by *Energy Future*, a report of the Energy Project at the Harvard Business School (Stobaugh and Yergin 1979).

In many ways this was an excellent major study. On oil it highlighted the growing dependence of the US on imports from the Middle East, and correctly pointed to the US production peak as the precursor for the 1973 oil price shock

(writing: "*In 1968, the State Department sent the word to foreign governments—American oil production would soon reach the limits of its capacity. Friendly governments needed to know that the cushion of the U.S.'s extra capacity, which could be called into production during an emergency, was about to disappear.*")

And on the US' other 'mainstream' fuels, of gas, nuclear and coal, the report correctly pointed to difficulties that would make fully offsetting the US losses in domestic oil production difficult. Moreover, the report highlighted the need for increased work on all the 'alternative fuels', of which energy conservation was seen as a major opportunity.

But on global oil, just as with the Carter speech, the report got the timing of the global oil supply peak wrong, despite the technical forecasts set out above in Chap. 3.

For example, the authors write (p9 ff): "*At best, enhanced recovery and other unconventional means can help to keep domestic [oil] supplies flowing at current levels. To be sure, outside the United States new oil fields in Mexico and China are important and will augment world supplies, but, claims to the contrary, they are unlikely to make a substantial change in the world oil balance.*" The authors thus wrote (p4): " ... *higher real oil prices seem assured for the future, with the only questions being how soon and how high*". Indeed, the report's Chap. 1, from which these quotes are taken, is titled: 'The End of Easy Oil'; with a sentence on p13 emphasising: "*The easy days of easy and cheap oil are truly over ....*"

Moreover, it is important to note that the oil view in *Energy Future* was not just that of a few researchers, but reflected the general view at the time. In the Acknowledgments the authors write: "*In the course of researching and writing this book, we two communicated with over three hundred business executives, government officials, labor union leaders, analysts, academics, and other specialists ... Our co-authors had similar exchanges with many hundreds more.*"

Therefore, not surprisingly, it was a good report overall, with many conclusions and recommendations valid still today. But on the specific issue of future global oil supply, just as with President Carter's speech, the question arises of the disconnect between the view of future oil supply in the report and the global forecasts available at the time from reputable organisations listed above in Chap. 3 (i.e., forecasts based on the estimated global URR for conventional oil, and which indicated therefore that global production of conventional could keep rising until about the year 2000, at which point its peak would be reached).

Was it that those advising President Carter, this Harvard study, and society in general were not aware of these forecasts; or were these forecasts discounted for some reason? In this context it may be indicative that *Energy Future* report does not seem to reference Hubbert.

Overall, it would seem that a solid explanation still awaits for the large 1970s/80s disconnect between the general expectation at that time of future global oil supply ('running out soon') and the technical URR-based forecasts from a range of recognised authorities (of: 'no peak in conventional supply till ~the year 2000, and large resources of non-conventional oil potentially available after that'). If readers have insight into this conundrum, I would be pleased to hear.

9. *Motivation for Petroconsultants Data, and the Data Collection Process*

Campbell notes that the oil majors have long been legally restricted by competition law from talking to each other, but naturally have wanted to know about the geology and discoveries being made in basins of interest. The Petroconsultants database was thus a useful medium for this, where the oil companies supplied data for fields they knew, in exchange for accessing data on the fields of others.

Note that in the early years the Petroconsultants data were mainly from insider industry sources known personally to the relatively small number of people at Petroconsultants, such as Dr. George Leckie, who were entering field URR data into the database on a multi-year rolling basis. Later, with the purchase of Petroconsultants by IHS Energy, some of the more mature analysts of the topic speculate that many of these insider contacts have been lost, and the data less reliable as a consequence. A second key factor here also is the proliferation of small independents, and also now of control of most of the world's oil by national oil companies. Both factors undoubtedly make it far harder than previously for data-assembly ('scout') companies to collect the valid oil reserves data that oil forecasting requires.

10. *Minor Error in 'The End of Cheap Oil'*

In another place in the text of *The End of Cheap Oil* the reference should be to *conventional* oil only, where this is clear from the remainder of the article.

11. *IHS Energy Definitions*

The text in the paper gives data for end-2011, of 2486 and 1248 Gb, for global 'Liquids' (IHS Energy definition) cumulative discovery and production, respectively. Though the categories of oil included are not fully clear, very probably they are the same as in the Figures in that article, and hence these latter two data points are included in Fig. 2.23.

12. *Craig Bond Hatfield*

By contrast, there have been good articles on peak oil in *Nature*, going back for example to Craig Bond Hatfield's *Oil Back on the Global Agenda*, Nature, Vol. 387, page 121; May 8 1997. (For an excellent insight into Hatfield's views on peak oil, see: *The Cassandra of Toledo: A Requiem For Mitigation* posted by on the Oil Drum website 'Prof. Goose' on March 14, 2007–12:15 pm, see http://www.theoildrum.com/node/2364.)

13. *Trying to Get BP to Add a Strong Caveat to the Proved Reserves Data in Their 'Stats. Review'; and Also on the Use of the R/P Ratios Listed*

At Reading University we made an effort to get BP to include a caveat on the proved reserves data in its otherwise excellent *Statistical Reviews*. We wrote to BP's Chairman to explain why this caveat was needed, and cautioning against just passing the request to Peter Davies. Back came a response from Davies saying that no such caveat was needed. (A corresponding caveat was, and is still, also needed cautioning against drawing any conclusions about security of future oil supply based on R/P ratios.)

14. ***Monitoring Errors in Past Oil Forecasts***

Hallock et al. (2014) in Sect. 4.7 of their paper write: "*It is surprising that institutions producing mid- to long-term forecasts of oil production, prices, etc., for use by governments and investors, do not periodically test the performance of their models or their past predictions. Such performance reviews would indicate whether underlying model assumptions and methods are valid or in need of change.*"

It is indeed surprising. On one occasion when we asked an individual at the IEA about the rationale for certain findings in a fairly recent *WEO* forecast, the reply was roughly: 'all that modelling has now gone, no-one today knows how those results were arrived at.' Given that the IEA's *WEO* was described by one senior UK government energy official as "the bible" it is important that past successes and mistakes by organisations such as the IEA be documented for future reference.

And there is a wider lesson here about the need for all organisations on which society depends to retain knowledge, despite staff turnover or other pressures. As others have noted, it is all too common across government departments of every hue, and also in the wider society, for lessons learned from previous situations (say, an outbreak of foot-and-mouth, or responses to an earthquake) to be no longer available when the situation recurs.

15. ***Illinois' Third Oil Production Peak***

For example, the extensive glacial drift in Illinois which generated a small pre-seismic exploration production peak and a larger post-seismic one; and the unexplored Middle Ordovician rocks which led the Illinois State Geological Survey to estimate a 4 Gb ultimate, compared to cumulative production of 1.8 Gb. From this extra oil Hubbert correctly predicted a third peak for Illinois (Hubbert 1956 p13; see also Laherrère 2004).

16. ***Hubbert Did Not 'Get Lucky': On the data available to Hubbert, a US Lower-48 conventional oil peak (ex-NGLs) around 1965–1970 was essentially certain.***

On Hubbert's forecast for the US peak, Kaufmann and Cleveland (2001) wrote "In effect, Hubbert got lucky" in his predicted date for the US peak. I have not looked at this argument closely. The exact date of peak was certainly influenced by factors such as changes in US pro-rationing rules, the oil price and so on, that Hubbert could not foresee. But he was correct when he said that given the constraints of the amount of oil the US oil had produced by 1956 and the likely Lower-48 conventional oil URR quantities (ex-NGLs) that the industry was estimating, he had very little freedom to draw realistic future production profiles that did not peak around the dates he indicated; about 1965 for the 150 Gb URR estimate, and around 1970 for the 200 Gb estimate. This is easy to verify by a simple diagram, see, for example, Fig. 5.1 in Hall and Ramírez-Pascualli (2013).

17. *Impact on Forecasts of the IEA's Resources to Reserves 2005 report*

Of particular note was the IEA's publication *Resources to Reserves*, IEA, 2005. This had a chart, similar to Fig. 2.24, estimating the global quantity of recoverable oil by category versus its price range. Critically, this chart included a red horizontal bar to indicate the *total amount of oil required* globally by 2030 to meet the IEA's *World Energy Outlook* Reference forecast, of about 1000 Gb. Because this was much less than the total remaining recoverable oil resource shown on the plot (of about 4600 Gb) it supported the conclusion that "there are plenty of oil resources to meet foreseeable demand". (The oversight was to ignore those resources shown on the figure that were already in resource-limited decline, and those that would soon be.)

18. *The IEA's Understanding of Peak*

Campbell notes (personal communication): "*It might be worth mentioning that behind the scenes the IEA did come to understand and obliquely confess* [to peak oil] *in its 1998* World Energy Outlook, *with the famous table showing that demand rose to 111.5 Gb by 2020 but* [where this] *relied on 19 Gb of "Unidentified Unconventional" coming in before 2020.* [I was called] *in the evenings* [by a contact at the IEA because] *the subject had become so political that* [they] *could not be seen to be in touch with me.*"

However, see End-note 20, below on why the IEA subsequently changed its opinion back to its earlier position of 'no near-term global peak'.

19. *See also: David Fleming: The next oil shock?*

Prospect magazine, April 20, 1999.

20. *The IEA's Understanding of Peak, contd.: Submission to the UK House of Lords Committee (and ODAC's Submission, also).*

Though Priddle had at some point said to me, in effect, that 'peak is not difficult to understand, look at the US', the submission by the IEA to the House of Lords Select Committee in November 2001 by Priddle and Appert clearly set out the IEA's new view, in part strongly influenced by the then-recent USGS year-2000 assessment.

To get the full context, the whole section of UK House of Lords (2002) page 117 ff ('Examination of Witnesses') should be read. But the following are indicative extracts:

> – *Firstly, we think the resource base is ample, whether we are talking about oil or gas or coal or renewables or uranium. ... That is a slight change of emphasis from last year even and certainly from the 1998* World Energy Outlook *when we were more concerned about the possibility of resource constraints.*

> – *The greatest work in depth on the resource question has been done by the US Geological Service.* [Their year-2000 Assessment underpins] *there being ample resources available to meet the foreseeable demand for any fuel for the next 20 years.*

– *... we have taken a relatively conservative decline rate of 5 per cent per annum. ... As you can see, by the end of the period 60 Mb/d* [globally] *of replacement or additional production of oil is needed.* [A calculation then follows that if the investment needed to develop new oil averages $10 billion per 1 Mb/d of added production, then the 60 Mb/d of new oil required would call for an investment of $600 billion: figures that "*do not frighten the financial sector*". The error here was not to realise that dearth of past discovery of oil in fields meant that many of the oil regions that the IEA had in mind to produce the extra 60 Mb/d were already in solid decline, or close to such decline, such that in these regions $10 billion could not add 1 Mb/d of production. Moreover, the oil in the Middle East had questions of access, and also of correct volumes. As a result of not allowing sufficiently for these factors, it is not surprising that the IEA's view on oil price was:]

– (Mr. Priddle): *I am usually asked what I think about tomorrow's oil price and I always refuse to answer but for the purpose of making projections ...* [we have taken] *an assumption that the* [oil] *price on average would be $21 a barrel through the decade to 2010 and then rising in the next decade to $28 at the end of the period.*" [Note that the US' EIA's oil price projection at about 2001 was similar to this.]

The Select Committee heard from many other organisations, including BP, Total, ExxonMobil, Energywatch, the Int'l. Association of Oil & Gas Producers (OGP), and the European Petroleum Industry Association. (EUROPIA). The views expressed by BP and ExxonMobil to the Committee are given in End-notes below, while EUROPIA wrote (p 300):

... we also believe that:—The short-term volatility of oil prices between 1998 and 2001 does not signal a new energy supply crisis.

This House of Lords Select Committee report (Paper 82) thus provides a good summary of the government, organisational, oil company and consultancy views at that time on the future of global oil supply (and of the other energies). [It could make a useful Master's project to study and summarise these views, given that far from the oil price rising to '$21/bbl by 2010, and to $28/bbl by 2020' as the IEA had assumed, by 2008 the price had risen to an average of over $100/bbl, and almost certainly helped in part to trigger the 2008 recession.]

21. *Very little Global Oil Supply Research within the Oil Companies*

For example, I was told by a senior person in one oil major to the effect that: 'Oh, I think we have a lady back in head office looking at that sort of thing'.

This relates to the wider issue, mentioned elsewhere, of how dramatically few people around the world were looking quantitatively at global future oil supply, given the obvious importance of the topic. Even the teams in the IEA, EIA and OPEC, for example, have all been small for the size of the task; and also often suffered from a relatively high staff turnover such that key expertise became lost.

22. *ExxonMobil's view to the House of Lords Select Committee*

ExxonMobil, in a written memorandum to the House of Lords Select Committee mentioned earlier (UK House of Lords 2002, p 287), wrote:

> *Commentators tend to focus on reserve to production ratios as the measure of resource availability ... 40 years for oil and 61 years for gas. However, such ratios fail to take account of probable reserves and reserves yet to be discovered. In addition the development of new technologies will make ... heavy oils of South America and tar sands of Canada economically viable. Beyond that emerging technologies will render [GTLs] economically viable. On this basis, R/P ratios in excess of 100 years will likely be achieved. Resource availability is simply not the issue.*

23. *BP's Peter Davies' Response, at the House of Lords Select Committee, to the ODAC / University of Reading 'Oil Group' View*

The Oil Depletion Analysis Centre (of which I was Co-ordinator, and sole employee) submitted a Memorandum to this Select Committee (HoL Report, pp 39–40). This started with:

> 1. *The world faces almost certain near-term hydrocarbon supply shortages. The reasoning is as follows. ....*

Dr. Rayner Mayer, of The University of Reading 'Oil Group', also submitted evidence (pp 40–41), which started:

> 1. *The world's fossil fuel resources are limited and there is substantial evidence that the physical limits to supply of oil and gas are being reached. Oil supply will peak first possibly within five years to be followed by gas within a further 10 years. Thereafter the gap between supply and demand will steadily increase as the supply declines.*

Both of us also gave oral evidence in response to examination, pp 41–48.

At a later date, BP's Peter Davies also gave evidence. The company had submitted a Memorandum (pp 71–78), and Davies was examined orally ("Examination of Witness. Professor Peter Davies, Chief Economist, BP"; pp 78–82). It was during this oral evidence that the following very telling exchange took place:

- Lord Chadlington: "*Professor, we have had evidence from the Oil Depletion Analysis Centre that world oil and gas supplies will shortly start to decline. I wonder if you could comment about their findings and indicate how much credence and credibility you give to it?*"
- Answer: "*This is an issue which has been going on for over 100 years. I think in this very building in 1860 there was a concern about the global supplies of coal which was going to run out, and the issues on oil and gas are just the same. There is always a series of geologists who are concerned about imminent depletion of world supplies, they have been wrong for 100 years and I would be confident they will be wrong in the future. As of today there are very adequate supplies of oil and gas. With the proven reserves and today's technology there is 40 years of oil and 60 years of gas plus. I think there is every reason to believe that technology will continue to exceed the forces of depletion and we have continually replaced both oil and gas reserves over the past 30–40 years, in fact, we can go back even further. I would be very confident on the availability of fossil fuels for the foreseeable future.*"
- Lord Chadlington: "*Pretty clear, I must say. Not much doubt.*"

24. Question by Laherrère on the Volume of Methane Hydrates

Laherrère, based on early JOIDES exploration drilling results, was dubious that the 'bottom simulating reflector' ('BSR') line that shows up in ocean seismic indicated *very large volumes* of methane hydrates below this line, as was commonly supposed. (Remark at a conference on 'Peak Oil' at Imperial College, London, organised by the Oil Depletion Analysis Centre, 2001).

25. *Asymmetry in the Size of Oil Fields*

In terms of the asymmetry in the size of oil fields, the extent of this often surprises even those working in the area. For example, the Executive Summary of the UKERC *Global Oil Depletion* report (Sorrell et al. 2009) noted that: "*Although there are around 70 000 oil fields in the world, approximately 25 fields account for one quarter of the global production of crude oil, 100 fields account for half of production and up to 500 fields account for two thirds of cumulative discoveries.*"

26. *The History of Oil Discovery and Production*

It is fascinating, and moving, to contemplate the almost unimaginable amounts of human physical and intellectual labour, and also risk, drama and tragedy, that lie behind a chart such as Fig. 2.23. The chart reflects in a few simple curves over a century-and-a-half of focussed human endeavour, and many millions of individual stories. Yergin's two outstanding books on the history of oil are key to unlocking the passionate and heroic realities behind these cold data.

References

In addition to the references below and the data sources listed in Annex 2, there are many other good sources of information on oil production and endowment. These include ASPO conference papers, data from 'The Shift Project' (http://theshiftproject.org), and the resource assessments of hydrocarbon resources by, for example, France's IFP, or of all fossil fuels by Germany's BGR (www.bgr.bund.de).

There are also other very useful websites with a wealth of important detail and exposition related to peak oil, including that by Ron Swenson (the first website on the topic), the Oil Drum, ASPO Newsletters (discontinued but still available), ODAC Newsletters, David Strahan's '*Last Oil Shock*' (http://davidstrahan.com), and the websites of Matt Mushalik (Crude Oil Peak; http://crudeoilpeak.info), Euan Mearns (http://euanmearns.com), ASPO, Uppsala University (website being updated), ASPO France (http://aspofrance.viabloga.com), and ASPO-USA (http://peak-oil.org).

Adelman, M. A. (1995). *The genie out of the bottle—World Oil since 1970*. Cambridge (USA) and London: The MIT Press.

Adelman, M. A. (1997). Comment at IEA meeting, Paris, 1997, held to discuss future oil supply; see Campbell (Ed.) (2011), Chapter 4. (Other chapters also cover this meeting).

Aguilera, R., Eggert, R. G., Lagos, G. C., & Tilton, J. (2009). Depletion and the future availability of petroleum resources. *The Energy Journal, 30*(1), 141–174.

And see also

Aguilera, R., Eggert, R., Lagos, G. C., & Tilton, J. (2012). Is depletion likely to create significant scarcities of future petroleum resources? In R. Sinding-Larsen & F. W. Wellmer (Eds.), *Non-renewable resource issues: Geoscientific and societal challenges* (pp. 45–82). Germany: Springer.

Rogner, H., Aguilera, R., Archer, C., Bertani, R., Bhattacharya, S., Dusseault, M. B., et al. (2012). Energy resources and potentials. In T. B. Johansson, A. Patwardhan, N. Nakicenovic, & L. Gomez-Echeverri (Eds.), *Global energy assessment: Toward a sustainable future* (pp. 423–512). USA: Cambridge University Press.

Ahlbrandt, T. S. (2003*). Presentation on Iraq oil, Transportation Research Board Conference*, Washington DC.

Aleklett, K. (2012). *Peeking at peak oil*. New York: Springer. (Update in process).

Andrews, S., & Udall, R. (2015) Oil prophets: Looking at world oil studies over time. *The Oil Age, 1*(3), 41–61.

Attanasi, E. D., & Root, D. H. (1994). The enigma of oil and gas field growth. *AAPG Bulletin, 78*(3), 321–332.

Bardi, U., & Lavacchi, A. (2009). A simple interpretation of Hubbert's model of resource exploitation. *Energies, 2*, 646–661, ISSN 1996-1073, www.mdpi.com/journal/energies.

Bardi, U. (2011). *The limits to growth revisited*. New York: Springer.

R.W. Bentley, *Introduction to Peak Oil*, Lecture Notes in Energy 34,
DOI 10.1007/978-3-319-26372-4

Bardi, U. (2014). *Extracted—How the quest for mineral wealth is plundering our planet. A report to the club of Rome*. Vermont: Chelsea Green Publishing.

Barney, G. O. (Study Director). (1982). *The Global 2000 Report to the President: Entering the Twenty-First Century*. Penguin Books.

Bentley, R. W., Mannan, S. A., & Wheeler, S. J. (2007). Assessing the date of the global oil peak: The need to use 2P reserves. *Energy Policy, 35*, 6364–6382.

Bentley, R. W., & Boyle, G. (2008). Global oil production: Forecasts and methodologies. *Environment and Planning B: Planning and Design, 35*, 609–626.

Bentley, R. W., Miller, R. G., Wheeler, S., & Boyle, G. (2009). *Comparison of global oil supply forecasts*. UKERC Review of Evidence on Global Oil Depletion: Technical Report 7. (Also see Annex1: *Models of global oil supply for the period 2008–2030.*) (www.ukerc.ac.uk & search 'TPA reports').

Bentley, R. W. (2015a). Forecasting oil production using data in the BP Statistical Review of World Energy. *The Oil Age, 1*(1), 59–73.

Bentley, R. W. (2015b). A Review of some Estimates for the Global Ultimately Recoverable Resource ('URR') of Conventional Oil, as an Explanation for the Differences betwen Oil Forecasts. Part-1: *The Oil Age, 1*(3), 63–90. Part-2: *The Oil Age, 1*(4), 55–77. Part-3: To be submitted to *The Oil Age, 2*(1).

Bentley, R. W., & Bentley, Y. (2015a). Explaining the price of oil, 1861 to 1970: The need to use reliable data on oil discovery and to account for 'mid-point' peak. *The Oil Age, 1*(2), 57–83.

Bentley, R. W., & Bentley Y. (2015b). *Explaining the Price of Oil, 1971 to 2014: The need to use reliable data on oil discovery and to account for 'mid-point' peak.* Energy Policy, Special issue on '*The Enigma of Economic Growth*' Online publication: May 9, 2015, doi: 10.1016/j.enpol. 2015.04.028.

Brandt, A. (2009). UKERC Review of evidence for global oil depletion. *Technical Report 6: Methods offorecasting future oil supply*. July 2009: Ref.: UKERC/WP/TPA/2009/021. Energy and Resources Group, University of California, Berkeley.

BGR (1997). From: www.bgr.bund.de.

Campbell, C. (1991). *The golden century of oil 1950–2050—The depletion of a resource*. Dordrecht: Kluwer Academic.

Campbell, C. J. & Laherrère, J. (1998). *The end of cheap oil*. Scientific American, March.

Campbell/Uppsala (2005). See http://www.peakoil.net, or: Campbell, C. J. (2003). *The Essence of Oil & Gas Depletion*. Multi-Science Publishing, Essex, UK, pp 233–254.

Campbell, C. J. (Ed.). (2011). *Peak oil personalities*. Skibbereen, Ireland: Inspire Books.

Campbell, C. J. (2013). *Campbell's atlas of oil and gas depletion* (2nd ed). Springer, Berlin. [First edition (now out of print) was: Campbell, C. J., & Heapes, S. (2008). *An atlas of oil and gas depletion*. UK: Jeremy Mills Publishing Ltd.].

Campbell, C. J. (2015). Modelling oil and gas depletion. *The Oil Age, 1*(1), 9–33.

Chu, S., & Majumdar, A. (2012). Opportunities and challenges for a sustainable energy future. *Nature,488*, 294–303.

Cole, H. S. D. et al., (Eds.) (1974). *Thinking about the future: A critique of 'the limits to growth*. Science Policy Research Unit, Sussex University, Chatto & Windus.

Dale & Benson. (2013). Energy balance of the Global Photovoltaic (PV) Industry—Is the PV Industry a net electricity producer? *Environmental Science & Technology, 47*(7), 3482–3489. [See also: Prieto & Hall. *Spain's photovoltaic revolution—Energy return on investment*. Berlin: Springer. ISBN: 978-1-4419-9436-3xx.)].

Deffeyes, K. S. (2001), & (Paperback: 2003). Hubbert's Peak. USA: Princeton University Press. See also: Deffeyes, K. (2005). *Beyond Oil*. Hill and Wang.

EIA. (2013). US Energy Information Administration report: '*Technically recoverable shale oil and shale gas resources: An assessment of 137 shale formations in 41 countries outside the US*'. Update June 13, 2013. [Incidentally, it makes *no sense* to quote such uncertain data to this precision; it is done so here only that the original source of the data can be identified.].

Ehrlich, P. R., Ehrlich, A. H., & Holdren, J. P. (1977). *EcoScience: Population, resources, environment* (pp. 400–404). San Francisco: W.H. Freeman.

Energyfiles Ltd. (2005). Personal communication.

Gagnon, N., & Hall, C. A. S. (2009). A preliminary study of energy return on energy invested for global oil and gas production. *Energies, 2*, 490–503.

Goldsmith, E., et al. (1972). *A blueprint for survival*. London: Penguin

Grubb, M. (2001). Who's afraid of atmospheric stabilisation? *Energy Policy, 29*(11), 837–845.

Guilford, M. C., Hall, C. A. S., O'Connor, P., & Cleveland, C. J. (2011). A new long term assessment of energy return on investment (EROI) for U.S. Oil and Gas Discovery and Production. *Sustainability, 3*(10), 1866–1887. doi:10.3390/su3101866.

Hall, C. A. S. (2008). Why EROI matters. EROI on the web: (Part 1 of 6). *The oil drum*. April 1, 2008. http://www.theoildrum.com/node/3786.

Hall, C. A. S., & Day, J. (2009). American Scientist, 97, 230–237. [Gives EROI of PV as ~8; value of 2.5 is from Prieto & Hall, *Springer Briefs in Energy*, 2013].

Hall, C. A. S., Balogh, S., & Murphy, D. J. R. (2009). What is the minimum EROI that a sustainable society must have? *Energies, 2*, 25–47.

Hall, C. A. S., & Ramírez-Pascualli, C. A. (2013). *The first half of the age of oil. An exploration of the work of Colin Campbell and Jean Laherrère*. Berlin: Springer.

Hall, C. A. S., Lambert, J. G., & Balogh, S. B. (2014). EROI of different fuels and the implications for society. *Energy Policy, 64*, 141–151.

Hallock, J. L. Jr., Wu, W., Hall, C. A. S, & Jefferson, M. (2014). Forecasting the limits to the availability and diversity of global conventional oil supply: Validation. *Energy, 64*, 130–153. http://www.sciencedirect.com/science/journal/03605442/open-access.

Höök, M., & Tang, X. (2013). Depletion of fossil fuels and anthropogenic climate change—A review. *Energy Policy, 52*, 797–809.

House of Lords, UK (2002). House of Lords Select Committee on the European Union considering the EU's Green Paper (Nov. 2000) *Towards a European strategy for the security of energy supply*. Select Committee's report: *Energy supply: How secure are we?*, Session 2001-02 14th Report, HL Paper 82, published 12th February, 2002.

Hubbert, M. K. (1938). *Technocracy research library*. Accessed March 28, 2014, from http://www.technocracy.org/technical-alliancetn/17-m-king-hubbert/205-hubbert.

Hubbert, M. K. (1956). Nuclear energy and fossil fuels. In *Drilling and production practice* (pp 7–25). The American Petroleum Institute, 1956 (Note that for convenience we refer to the 'Hubbert 1956' method. This was clearly developed over a number of years, and Hubbert refers to earlier publications in 1948 and 1950. Similar remarks apply to the dating given for Hubbert's later works.).

Hubbert, M. K. (1962). *Energy resources: A report to the committee on natural resources of the national academy of sciences—National Research Council*. Washington DC: Publication 1000-D.

Hubbert, M. K. (1967). *Degree of advancement of petroleum exploration in United States*.

Hubbert, M. K. (1982). Techniques of prediction as applied to the production of oil and gas. In *Oil and Gas Supply Modeling: Proceedings of a Symposium at the Department of Commerce* (pp 16–141), Washington DC, June 18–20 1980. [Report very kindly supplied to me by Jens Junghans.].

IEA. (2005). *Resources to reserves*, 2005 (or just possibly, 2008) edition.

IEA. (2013). *Resources into reserves* (p 75). 2013 edition.

IIASA. (2012). *Global energy assessment study (GEA)—Toward a sustainable future*. Cambridge, UK and New York, NY, USA: International Institute for Applied Systems Analysis, Vienna, Austria, and Cambridge University Press.

Ivanhoe, L. F. (1996). Updated Hubbert curves analyze world oil supply. *World Oil,217*(11), 91–94.

Jackson, P. M., & Smith, L. K. (2014). Exploring the undulating plateau: The future of global oil supply. *Philosophical Transactions of the Royal Society of London A: Mathematical, Physical and Engineering Sciences, 372*, 20120491.

Jakobsson, K., Bentley, R., Söderbergh, B., Aleklett, K. (2012). The end of cheap oil: bottom-up economic and geologic modelling of aggregate oil production curves. *Energy Policy, 41*, 860–870.

Jefferson, M. (2012), Shell scenarios: What really happened in the 1970s and what may be learned for current world prospects. *Technological Forecasting & Social Change, 79*, 186–197.

Kaufmann, R. K. and Cleveland, C.J. (2001). Oil production in the lower 48 states; Ecological, geological and institutional determinants. *The Energy Journal*, 22, 27–49.

Kharecha, P. A., & Hansen, J. E. (2008). Implications of "peak oil" for atmospheric CO_2 and climate. *Global Biogeochemical Cycles, 22*

Kiely, J. (1980). World energy in the 21st century. *Fourth Wilson Campbell Memorial Lecture, reported in CME, IMechE., UK*, May 1980, pp 26–32; commenting on World Energy Conference (now World Energy Council) Commission report: *World energy: Looking ahead to 2020* . This said "The French Petroleum Institute prepared the Petroleum Resources and Production study".

Klett, T. R., Gautier, D. L., & Ahlbrandt, T. S. (2005). An evaluation of the U. S. Geological Survey world petroleum assessment 2000. *AAPG Bulletin, 89*(8), 1033–1042.

Laherrère, J. H. (2000). *Is the USGS 2000 assessment reliable?*www.oilcriss.com/laherrere/usgs2000.

Laherrère, J. H. (2004). Comparing Hubbert's data and data to-date, Illinois shows a small fourth peak, apparently linked to the high oil prices of late-70s/early-80s. www.oilcrisis.com.

Laherrère, J. H. (2011) Presentation. *Backdating is the key*. http://aspofrance.viabloga.com/files/JL_ASPO2011.pdf.

Laherrère, J. H. (2014). *Notes for ESCP Roundtable*. London, 17 February, 2014; a discussion workshop of the *Energy Policy* journal special issue on x.

Laherrère, J. H. (2015). A global oil forecasting model based on multiple 'Hubbert' curves and Adjusted Oil-Industry '2P' discovery data: background, description & results. *The Oil Age,1* (2), 13–37.

Lambert, J., Hall, C.A.S., Balogh, S., Gupta, A., & Arnold, M. (2014). Energy, EROI and quality of life. *Energy Policy, 64*, 153–167. http://authors.elsevier.com/sd/article/S0301421513006447

Mackay, D. (2009). *Sustainable energy: without the hot air.* UIT Cambridge. ISBN 0-9544529-3-3.

Maugeri, L. (Chief Economist of ENI). (2004). Oil: never cry wolf—Why the petroleum age is far from over. *Science, 304*, 1114–5; and see also the rebuttal in Meng and Bentley (2008).

Meadows, D. H., Meadows, D. L., Randers, J., & Behrens, W. W. III (1972). *The limits to growth*. New York: Universe Books, Potomac Associates.

See also:

- Meadows, D. H., Meadows, D. L., & Randers, J. (1992). *Beyond the limits*. VT: Chelsea Green.
- Meadows, D. H., Randers, J., & Meadows, D. L. (2004). *Limits to growth: The 30-year update*. VT: Chelsea Green.

Meng, Q. Y., & Bentley, R. W. (2008). Response to L. Maugeri: Why the world should not count on abundant supplies of oil. *Energy, 8*, 1179–1184.

Miller, R. G. (2012). Reserves uncertainties: Fallow fields (and other questions). In *ASPO Conference, Vienna, May 2012*. http://www.aspo2012.at/wp-content/uploads/2012/06/Miller_aspo2012.pdf.

Miller, R. G., & Sorrell, S. (2014). The future of oil supply. *Philosophical Transactions of the Royal Society of London A: Mathematical, Physical and Engineering Sciences,372*, 20130179.

Miller, R. G. (2015). A bottom-up model of future global oil supply. *The Oil Age, 1*(2), 39–55.

McGlade, C., & Ekins, P. (2015). The geographical distribution of fossil fuels unused when limiting global warming to 2 °C. *Letter to Nature, 517*, 187–190.

McGlade, C. (2015). An introduction to the bottom-up economic and geological oil field production model 'BUEGO'. *The Oil Age*, 1(3), 9–40.

Mills, R. M. (2008). *The myth of the oil crisis*. Preager. (See Figure 2.24 for the IEA's view of the likely global contribution from CO_2-EOR.).
Moore, C. L. (1971). In Cram, I. H. (ed.). *Future Petroleum Provinces of the United States—Their Geology and Potential*. American Association of Petroleum Geologists, Memoir 15.
Muggeridge, A., Cockin, A., Webb, K., Frampton, H., Collins, I., Moulds, T., & Salino, P. (2014). Recovery rates, enhanced oil recovery and technological limits. *Philosophical Transactions of the Royal Society of London A: Mathematical, Physical and Engineering Sciences,372*, 20120320.
NPC: National Petroleum Council (US). (2007). *Facing the hard truths about energy—a comprehensive view to 2030 of global oil and natural gas.*
Odell, P. (1997). Comment at IEA meeting, Paris, 1997, held to discuss future oil supply; see Campbell (Ed.) (2011), Chapter 4.
Perrodon, A., Laherrère, J. H., & Campbell, C. J. (1998). *The world's non-conventional oil and gas*. The Petroleum Economist Ltd, London, March 1998; 113 p; see http://textor.com/cms/dPEWNC.html.
Petroconsultants Reports (1994–1996).
– *Undiscovered Petroleum Potential*. (1994). J. H. Laherrère, A. Perrodon and G. Demaison. Petroconsultants S.A., Geneva, 383 pp.
– *The World's Supply of Oil, 1930–2050*. (1995). C. J. Campbell and J. H. Laherrère. Petroconsultants S.A., Geneva, Oct. 1995, 650 pp; and on CD-ROM.
– *The World's Gas Potential*. (1996). J. H. Laherrère, A. Perrodon and C. J. Campbell. Petroconsultants S.A., Geneva, July/August 1996, 200 pp; and on CD-ROM.
PFC Energy. (2005). *Presentation at the 'global oil depletion debate'*. London: Energy Institute, also personal communications.
Pogue, J., & Hill, K. (1956). *Future growth and financial requirements of the world petroleum industry*. Chase Manhattan report, presented at the Annual Meeting of the Petroleum Branch of the American Institute of Mining, Metallurgical and Petroleum Engineers.
Porter, E. D. (1995). *Are we running out of oil*? American Petroleum Institute, Discussion Paper 081.
Radetzki, M. (2010). Peak oil and other threatening peaks—Chimeras without substance. *Energy Policy,38*(11), 6566–6569.
Rühl, C. Blog titled 'Peak Oil', dated February 27th 2014. Accessed May 28, 2014, at: http://www.bp.com/en/global/corporate/about-bp/energy-economics/energy-blog.html.
Shell (1979). (A.F. Beijdorff). Energy Efficiency, Group Planning, Shell International Petroleum Company, London, April 1979; p 1.
Smith, M. R. (2015). Forecasting oil and gas supply activity. *The Oil Age, 1*(1), 35–58.
Sorrell, S., Speirs, J., Bentley, R., Brandt, A., & Miller, R. (2009). Global oil depletion—an assessment of the evidence for a near-term peak in global oil production. *Technology and Policy Assessment report: UK Energy Research Centre, August 2009*. ISBN 1-903144-0-35 (Go to: www.ukerc.ac.uk & search 'TPA reports').
Sorrell, S., & Speirs, J. (2009). UKERC Global Oil Depletion report; Technical Report 1 (from TPA reports, www.ukerc.ac.uk) *UKERC Review of Evidence on Global Oil Depletion. Data Sources and Issues*. July 2009: REF UKERC/WP/TPA/2009/016.
Sorrell, S., & Speirs, J. (2014). Using growth curves to forecast regional resource recovery: approaches, analytics and consistency tests. *Philosophical Transactions of the Royal Society of London A: Mathematical, Physical and Engineering Sciences, 372*, 20120317.
Stobaugh, R., & Yergin, D. (Eds.). (1979). *Energy future—report of the energy project at the harvard business school*. New York: Random House.
Thompson, E., Sorrell, S., & Speirs, J. (2009). UKERC review of evidence on global oil depletion technical report 3. *The Nature and Importance of Reserve Growth July 2009: REF UKERC/WP/TPA/2009/018.*

UK Department of Energy, Marshall, W. (1976). *Energy research and development in the United Kingdom*, Energy paper No. 11, UK Department of Energy, p. 12.

US DOE (US Department of Energy) report. (1990). *USDOE/EIA -0534 (1990) US oil and gas reserves by year of field discovery*, August.

USGS, Ahlbrandt et al., T. (2000). U.S. Department of the Interior. U.S. geological survey world petroleum assessment 2000—Description and results. U.S. Geological Survey Digital Data Series - DDS-60. (Available on CD-ROM; go to: http://pubs.usgs.gov/dds/dds-060).

Wang, J., Feng, L., Tang, X., & Höök, M. (Submitted for publication). *The implications of supply constraints of fossil fuels for climate change: a supply-side analysis.*

Ward, B., & Dubois, R. (1972). *Only one earth: the care and maintenance of a small planet*. UK: Penguin Books.

Ward, J. D., Mohr, S. H., Myers, B. R., & Nel, W. P. (2012). High estimates of supply constrained emissions scenarios for long-term climate risk assessment. *Energy Policy, 51*, 598–604.

Warman, H. R. (1979). *Oil Crisis ... Again?* British Petroleum Ltd.

Watkins, G. C. (2006). Oil scarcity: What have the past three decades revealed? *Energy Policy, 34*, 508–514.

Wold, E. (2015). Bottom-up modelling of future global oil supply and associated spend. *The Oil Age, 1*(4), 19–39.

World Bank. (1981). Global energy prospects. *World Bank Staff Working Paper No. 489*, pp. 37, 46.

Yergin, D. (1991). *The prize: The epic quest for oil, money, and power.* Simon and Schuster.

Zapp, A. D. (1962). Future petroleum producing capacity of the United States. *US Geological Survey* Bulletin, *1142-H*, 36.

Index

Note: The letter "n" following page numbers refer to endnotes.

R.W. Bentley, *Introduction to Peak Oil*, Lecture Notes in Energy 34,
DOI 10.1007/978-3-319-26372-4